Heinrich-Karl Podszeck

Carrier Communication over Power Lines

Fourth Revised Edition

Springer-Verlag Berlin · Heidelberg · New York 1972

HEINRICH-KARL PODSZECK

Chief Engineer of the Power Line Carrier Section Siemens Aktiengesellschaft,
Munich, Germany

With 88 Figures

ISBN-13: 978-3-642-46288-7   e-ISBN-13: 978-3-642-46286-3
DOI: 10.1007/978-3-642-46286-3

Preface to the Fourth Edition

Carrier communication over power lines, like carrier traffic over postal lines, has a tradition of about 50 years. It has become a vital element in electric system operation. Being restricted to power networks, it has a comparatively limited scope. Nevertheless, great efforts are being expended on technical advancement to keep pace with the demands made of both power and communication engineering.

As in previous editions of this book, it is endeavored in the fourth edition to familiarize the man in the field with the fundamental concepts on which this technique and its practical application are based. The physical and electrical characteristics of the equipment involved are described in a general way. For the sake of clarity, the author has refrained from giving a detailed account or illustrations of special versions, especially as the differences between the equipment supplied by the various manufacturers are so small. This is due to the use of transistors instead of tubes and to the universally adopted plug-in module construction technique. In the load dispatching plants of power supply networks, telemetering systems are turning more and more to data engineering methods and equipment to lighten the task of network management. Consequently, at certain points in the text, topics concerning remote data processing are treated, emphasis being placed on transmission channels operating at a considerably higher speed than was usual 10 years ago.

In preparing this new edition, the author was able to draw on the wealth of experience gained in his dealings with CIGRE experts from many countries over the past 20 years.

The author endeavored to present the data in a form easily understood both by the power engineer and the communication engineer. A definition of pure communication engineering terms was not felt to be necessary. An index has been provided to assist the reader. Wherever discussion of details would involve undue departure from the main subject, the reader is referred to pertinent publications or to explanations in the appendix. Figures in brackets refer to relevant publications.

The text is illustrated by photographs supplied by the manufacturers of the equipment, and certain passages are reproduced from original works with the permission of the publishers. The author extends his thanks to Dr.-Ing. E. ALSLEBEN and Ing. G. BERGMANN, both of Munich, who assisted him in preparing the fourth edition with contributions and valuable advice. Also, he is obliged to H. HEROLD, Munich, for doing the translation.

Munich, in January 1972   **H.-K. Podszeck**

Contents

List of Abbreviations

UCPTE   Union pour la Coordination de la Production et du Transport de l'Electricité

AIEE   American Institute of Electrical Engineers

CCITT   Comité Consultatif International Télégraphique et Téléphonique

CEI   Comission Electrotechnique Internationale

CIGRE   Conférence Internationale des Grands Réseaux Electriques a Haute Tension

IEC   International Electrotechnical Commission

1. Private Communication Networks for Electric System Operation

Supply of electrical energy to large areas through overhead transmission lines has been expanded to a point where an extensive communication network has become an indispensable aid in system management. The communication media are used for

a) electric system operation, in an effort to provide electric energy reliably and at low cost,

b) supervisory control of operating conditions, to satisfy the existing power requirements at the pre-arranged voltage and frequency,

c) limitation of power supply failure in extent and time.

To solve these problems, messages are exchanged in an interconnected network between the various power generating stations on the most appropriate operating schedule for steam plants and hydro-electric stations, and decisions are made as to the extent to which energy is to be generated locally or drawn from other sources. A certain schedule has in most cases been established for the transition points between the various sections of the compound network so that peak load demands can be economically met. Communication circuits are required also between the power generating stations and the bulk users (industry, municipalities, railways) to ensure a satisfactory power supply.

In the event of trouble in power transmission, speedy exchange of information is imperative also between widely spread stations of the power network so as to eliminate the trouble and to reduce the outage time to a minimum.

Thus, the communication network of a power company is called upon to connect the power plants, the transformer sub-stations, and the power transfer points to neighboring networks or bulk users. These "district communication networks", which serve the needs of a single power company, must be supplemented by long-haul circuits which extend beyond the local distribution range, to permit exchange of messages between all power plants pooled in the intermeshed network. The "long-haul network" interconnects only those offices which control the exchange of electrical energy within the entire area served by the interconnected power network.

1 Podszeck, Carrier Communication, 4th Ed.

These "load dispatcher" offices need not necessarily be accommodated in a power network station. They may as well be located in the administration building of a power company in a metropolitan area, far away from any power distribution facilities. All that is required is a communication system linking them with the stations which are of importance in their network as regards power generation and distribution. It is not only the telephone circuits which terminate in a dispatcher's control room. Mimic network diagrams display the switching condition and essential measurands, which are transmitted from the power network in uninterrupted succession by means of long-haul communication facilities.

The telephone is naturally the most important device for transmitting messages of a general nature. In some cases, however, such as the continuous remote supervision of measurands, the emphasis is on uninterrupted transmission without the presence of an operator. Other applications, selective line protection for instance, require the reporting over large distances of the operation of protective relays within fractions of a second, and lines must be disconnected without the aid of an attendant. These few examples show that in addition to telephony, other means of communications, such as systems for teleprinter message transmission, supervision, remote control, and selective line protection also play an important part.

To solve all these transmission problems, company-owned communication networks have been set up in the course of time. The public telephone network is not sufficiently free from trouble and is not always immediately available when an emergency arises. The delay times incurred in establishing connections, and dependence on external agencies in the event of trouble isolation and elimination are incompatible with the operational requirements of power supply systems. Besides, most of the stations are located far away from the densely populated areas covered by the public telephone network of communities, so that they would have to be connected by tie-lines to be specially laid for this purpose.

The freedom of alternative circuit patching and temporary interruption of circuits, which the postal authorities must insist on to permit routine testing and reconditioning of lines in an effort to ensure an equal grade of service for all subscribers, can hardly be brought into agreement with the necessity for continous transmission of measurands to a power station, where generator output is controlled with the aid of a criterion obtained through the summation of a number of variables transmitted over large distances. Even a brief interruption in the transmission of such a variable through the intervention of an external agency would falsify the basis for electric system operation.

All these considerations prompted the power companies to provide their own communication systems. In individual cases it might be advisable to use a circuit of the public network or to rent a line. This may be an economical solution where the distances involved are short. With increasing distances, the fees or rentals will soon reach a figure which is very high compared with the cost of installing and maintaining a company-owned system.

In larger cities even the municipal power companies tend to set up their own communication networks, to provide circuits not accessible to the general public, but at any time ready for the exclusive use by the power distribution agencies. Communication cables are here the rule, so that the circuits are protected against adverse weather conditions such as thunderstorms, high winds, and hoarfrost. The mode of operation is the same as that on normal postal lines.

In extensive power networks and in interconnected networks energy is supplied over distances much larger than those which have to be spanned in municipal communities. In the case of operating voltages ranging up to approximately 60 kV, the technical conditions are such as to permit the laying of a separate open-wire telephone pair on the tower structures of the transmission line. Where the power lines are run as underground cables, a communication cable is buried in the same trench. In either case, special protective measures must be taken at both ends of the communication line to prevent hazardous voltages between communication line and ground, induced by the power line, from reaching the communication equipment. Where more than one wire pair is required for communication, multi-pair communication cables are used instead of open-wire lines. In medium-voltage (10–60 kV) networks, these cables may be suspended from messenger wires attached to the towers, or the tensile strength of the cable sheath may be rated so that a messenger is not required. These "self-supporting" aerial cables are suspended below the power transmission lines and are used exclusively for communication purposes. However, they may also be designed as ground wires of overhead power lines. In this case they are no longer strung below but arranged above the power lines. As the regular ground wires, they are thoroughly grounded at the tops of the individual towers. The sheath of such a cable takes over the function of the ground wire of the power line system, while the interior accommodates several wire pairs for communication purposes.

The possibility of using ground wires designed as carrier frequency cables has also been examined. Cables of this type offer the great advantage of providing a large number of carrier frequency channels with only a few conductors. As a result of this, the weight of the cable can be kept low. Moreover, the undesirable leakage of HF communication currents into network sections where they are not required is avoided. This dissipation is quite a problem in intermeshed networks if the power lines are used also for message transmission. The available frequency range is not as narrow as it necessarily is in carrier communication over power lines, so that the standard carrier systems, developed for use by postal administrations can be employed also by power companies. Although these facts have been known for many years[1], it is only now that self-supporting carrier frequency cables of this type are used as ground wires of power transmission systems. These cables were uneconomical as long as a sufficient number of communication channels along a power line was not required or if distances were too long.

Similar to HT-influenced underground cables or open-wire lines, the wire pairs of self-supporting aerial cables must be terminated at their ends by devices which protect the communication equipment. While satisfactory results have been obtained with suitable protective devices on open-wire communication lines suspended from the towers of power lines carrying up to approximately 60 kV, underground cables and self-supporting tower-suspended aerial cables for voice frequency communication have been run in parallel with power systems carrying as much as 220 kV.

For the utility companies, the power distribution network proper is the most natural means of message communication because the stations to be interconnected are already connected through the power lines.

For technical and economical reasons, the underground power cables of the municipal power companies cannot be used for message communication. Considering the small distances to be covered within the area of a city, company-owned communication cables can be installed at much lower cost than a communication system using the municipal power distribution network as a message transmission medium. At increased distances between the communicating stations of a network, the provision of separate communication lines will no longer pay. Instead, special carrier frequency terminals are used to handle message traffic directly over the power transmission lines. This is technically feasible as long as the power lines are overhead transmission lines. Cables up to a length of approximately 20 miles may also be used for this purpose. Short power cable sections (not longer than 2 or 3 miles), interposed between overhead power lines, do not involve any technical difficulties worth mentioning.

For all power supply companies working over long overhead lines the power network itself is the most important means of communication. Power lines are much safer, mechanically, than open-wire communication

---

[1] See also 1st Ed. of this book [3, pp. 166].

lines and therefore afford the necessary reliability. As a result, carrier communication terminals have been extensively used in power networks carrying operating voltages of 60 kV and more. Their use is limited essentially by economic factors. At lower voltages, the networks are normally closely intermeshed and the individual stations are not separated by wide distances, so that special communication lines will be less expensive than coupling media, traps, and carrier terminals.

As long as the problem consists, essentially, in providing only a few channels over a long overhead power line to satisfy the operational and administrative requirements of the power plants, and this has been so far the case, carrier communication over the power network will remain the predominant communication method at the present state of communication engineering. Occasionally, however, the need arises for a larger number of communication circuits. Such large channel groups could be provided by means of multi-channel telephone systems, of the type used by postal authorities. These system must then be adapted to the specific operational requirements of power lines. The realization of these projects is frequently hampered by obstacles connected with frequency assignment and transmission range (see page 87). It proved to be more convenient so far to resort to radio communication in the VHF range, i. e. to construct radio relay connections. Directional antennas are here used for focusing the radiated energy in a certain direction and this permits operation at a comparatively low output power. The line of sight may be exceeded to some extent if the wave length is increased. This applies already to the VHF range. However, the transmission range cannot be extended in this way to any desired point. Increasing the output power would be an uneconomical solution to the problem. The most efficient measure consists in erecting the antenna at the most elevated point. Repeater stations are required where large distances must be covered. Nevertheless, the cost of such radio relay connections is high enough to restrict their use to multi-channel operation between the nodal points of a communication network.

Operation in the VHF range permits omni-directional radiation at a low send power so that initial costs can be kept within reasonable limits. This mode of operation lends itself particularly well to communication with mobile stations within a certain area, i. e. with line crews. The range of these communication systems, designed along the lines of police radio, also depends essentially on the height of the antenna. This line crew radio system is used to advantage in medium-voltage networks of limited size, where it serves, so to speak, as a mobile means of extending the existing wire-bound communication system. In the case of extra-high voltage power lines extending over large areas it can no longer be used, however, because of its restricted range. Omni-directional radiation in the VHF range between fixed-plant installations is unsatisfactory from a technical point of view. Moreover, the distances to be spanned even in normal medium-voltage networks carrying up to 30 kV are usually larger than the line of sight.

Carrier communication over power lines is the preferred means of communication where large distances have to be covered. Examinations as to which transmission medium, aside from the power network, might also present an economically and technically sound solution for a specific project, must rule out right from the outset all communication lines requiring separate supporting structures, because erection and maintenance costs would not be economically justified in the case of large distances. With operating voltages ranging up to 60 kV, i. e. in medium-voltage networks, open-wire lines or aerial cables suspended from the power line towers must be taken into consideration. In the case of operating voltages of 220 kV and higher, a radio link might be the other choice. There are no hard and fast rules providing a basis for a decision as to where company-owned communication lines, carrier communication over power lines, or radio relay connections will best serve the purpose. At any rate, they cannot be formulated precisely enough to permit an unfailing judgment in each specific case. Carrier communication over power lines is used to best advantage where the operating voltage is high and the number of communication channels small.

2. Communication Problems in Electric System Operation

The communication media used by the utility companies for operating their power distribution networks may be classified into three groups: telephone, supervisory control, and teleprinter systems. The oldest and most important means of communication is telephony. Beginning in 1920, various countries started to construct carrier telephone systems for operation over power lines. These initially individual connections were gradually integrated into larger networks. In the course of time, pooling of power distribution systems led to the construction of large company-owned carrier telephone networks. The interconected network combined in the UCPTE, for instance, includes thousands of carrier terminals.

In 1932 a number of specialized communication tasks for electric system operation were grouped under the heading "Fernwirken" (tele-operation) in German speaking countries. These included metering, counting, control, and supervision over large distances. The use of this generic term underwent changes in the course of time. Today this term is no longer bound up with the management of power distribution systems. It is now used in a much broader sense for the transmission and the processing of system-bound, not arbitrarily variable technical information, between man and technical facilities, or vice versa, or between technical facilities only. "Teleoperation" systems therefore include the facilities for input, transmission, and output of information as well as the associated information processing equipment[1]. Within the scope of electric power distribution the following systems range under the heading "teleoperation", in the following termed supervision and control systems:

1. Supervisory systems, such as metering, indicating and counting systems,

2. Remote control systems,
and combinations thereof.

The term "metering" implies the determination of an instantaneous value, such as active kW or reactive power kWr contrary to "counting", which covers quantities, such as electric work (kWhr). Measured values can be indicated or registered on a time basis by means of recording instruments on a paper tape which is advanced by a clock movement (ink recorders). Counting results, however, are registered by devices which indicate, write, or print quantities over a certain period (drum counters, maximum load recorders, printometers).

Quite generally, "telemetering" refers to the transmission of metered information if, for the purpose of long-distance transmission, the measurand is converted at the sending station into an auxiliary criterion, and thereby made independent of the length and other physical properties of the transmission line [5]. In "telecounting", a certain amount of work to be counted is represented by a current pulse, which is transmitted by the brief closure of a counting contact. This pulse is used to advance a counter in the receiving station by one step. If the quantity to be counted increases rapidly, the pulses arrive at frequent intervals and the counting mechanism is advanced at a high rate. If the quantity grows slowly, fewer pulses arrive per time unit and the counter is stepped at a slower rate.

The auxiliary intelligence, into which metered or counted information is transformed before transmission, may be generated by an "analog process" similar to the one just described. In metering applications, for instance, the measurand may be represented by a proportional pulse frequency or by a frequency varying in proportion to the measurand. In telecounting, each pulse transmitted corresponds to a certain quantity. If such an analog process is adopted, the fundamental difference between

---

[1] This definition was accepted in 1960 by representatives of Germany, Austria, and the German speaking part of Switzerland. A comparable all-embracing term does not exist in English speaking countries.

metering and counting must be clearly understood by the communication engineer. If telemetering pulses get lost as a result of interference along the transmission path, the instantaneous meter reading is wrong. As soon as the trouble is over, the indicating instrument (or the recording instrument) will provide the correct information. On the other hand, if telecounting pulses get lost (or if interference pulses are added), the counter in the receiving station is not correctly stepped and this error falsifies the counting result with no possibility of detecting the error. In the course of time the error becomes cumulative. Great efforts have therefore been made to develop a "counter reading transmission system" capable of detecting and transmitting counter readings at certain intervals. Counter reading transmission eliminates the influence of any transmission errors which might have occurred ahead of the brief reporting process.

Another form of proportional intelligence, used for conveying measurands or counter information over the transmission path, are coded signals. The measuring range is here divided into small sections and each section is assigned a coded signal (similar to the coding of letters in telegraphy). In telecounting, too, each pulse may be coded. These "pulse-code methods" offer a high measure of immunity against interference so that, contrary, to an analog process, the requirements imposed on the transmission channel with respect to dependability are no longer different for telemetering and telecounting.

Remote indication and signalization systems report operating conditions, mostly by way of signal contacts, such as switch positions, and the operation of ground fault detecting relays or other alarm facilities. Contrary to telemetering channels, the transmission channels employed for this purpose are occupied only for a brief period each time and are therefore available for many different indications if the so far almost exclusively used start-stop equipment is retained for coding, transmitting and receiving the information. Of late, however, advances scored in the development of electronic components, brought about a revival of time-division multiplex systems with cyclic information read-out, also for use in remote indication systems, so that the transmission channel is continuously engaged. Earlier time-division equipment, depending on electro-mechanical components, has been almost completely abandoned because of its lack of adaptability to the different conditions of the stations, its low cycle speed, and the rapid wear on moving parts.

Remote control systems [4, 9] have input and output equipment similar to that used for remote indication, or they may function as expanded telemetering systems, where the transmitted variable causes a control element to be continuously operated at the receiving end.

Network protection systems cannot be regarded as belonging to the scope of supervision and control if they merely contain detecting relays

Correction

On page 45, 8th line from the bottom, please
read: radio frequency instead of Hz;

4th line from the bottom, please
read: Hz instead of radio frequency.

On page 91, 8th line from the top, please
read: 114 instead of 144.

Podszeck, Carrier Communication, 4th Ed.

at both ends of the power line, i. e. relays which respond to overcurrent, ground fault, and reversal of energy transmission. Protective relays of this type may trip a circuit breaker merely in response to a local change of conditions. However, signal transmission between the relay sets at the ends of a line is additionally required for high-speed tripping independent of the point of fault. This applies in particular to networks carrying more than 110 kV, which have the neutral of the three-phase system solidly grounded. The automatic reclosure method is here adopted to eliminate a fault in the power line as quickly as possible, and a high-speed protection system is therefore indispensable. With signal transmission between the protective relay sets, a protection system comes within the scope of supervision and control [6].

The duties which the communication systems of power companies have to fulfill can be divided into two different categories. They may be categorized as to whether only one direction of the two possible directions of a communication line is used, or whether both directions are used one at a time, or whether both directions can be used simultaneously. They may as well be grouped according to transmission times. Some of them engage a circuit all the time. In other cases the holding time is limited arbitrarily by operators. Finally, limitation of the time a circuit is occupied may be inherent in the operating principle of the connected equipment. Limitation is then automatic.

If the more frequent transmission duties in electirc system operation are grouped according to the abovementioned scheme, an outline (Fig. 1) can be given which, however, cannot possibly cover all variants occurring in practical operation. Aside from the coding principles based on start-stop equipment, modern electronic telemetry depends, as previously mentioned, also on a time-division principle which holds one direction of transmission continuously engaged. A specific problem arises with selective metering. In most cases the metering points are selected from the receiving station. Obviously, two transmission channels are here required, one in the direction in which the metered data flows, and another one in the opposite direction for the selection process. Similarly, a transmission channel in the opposite direction of the actual remote control channel is required in remote control applications, to give the operator a continuous picture of the result obtained at the other end, and to permit him to react accordingly. Likewise, a switch position indicator system can only do with one direction of transmission if the necessity of interrogation from the receiving station is waived. Channels will be required in both directions of transmission, although not at the same time, if it should be possible for the operator in the supervising room to interrogate the position of the switches in the distant, supervised station.

Direction	Transmission task		Transmission period		
			Unlimited	Arbitrarily limited	Automatically limited
Both directions simultaneously	A	Telephony		Each call	—
	B	Telegraphy	—	Each message	—
	C	Remote supervision and control			
		Line protection	—	—	Any type of line protection
Both directions, one at a time		Indication	—	—	Switch position, Buchholz protection, pressure gage, manual reader pick-up
		Control	—	—	Switch position, synchronizing
		Metering		Selective metering	On-call telemetering
One direction only		Metering	1. Recorder 2. Summation of measurands at receive station 3. Derivation of control pulses at receive station		
		Counting	Each count	—	Counter reading
		Control	Automatic control of frequency, power, etc.	Continual control of generators, water shut-off valves, reference point indicators, etc.	Step-by-step control of tapped transformers

Fig. 1. Communication problems in electric system operation

Where a separate communication line is available, transmission equipment requirements do not differ for transmission in both directions, e. g. telephone conversation, and transmission in one direction, as in telemetering applications. If, on the other hand, a communication circuit is established with the aid of carrier terminals, each station must be equipped with a transmitter for the outgoing, and with a receiver for the incoming traffic, same as in radio communication. Transmission in both directions therefore requires twice the amount of equipment necessary for one-way traffic. When a carrier communication system is being planned, a decision must therefore be made whether one or both directions of traffic are required to solve a specific problem. In all cases where a communication channel is only briefly engaged in one direction, e. g. for switch position interrogation in a switch position indicator system, the cost of a separate interrogation channel is not economically justified. Economical and technical justification (frequency conservation) will normally exist only in connection with other transmissions flowing in the same direction.

It is common practice to install carrier telephone terminals which permit operation in both directions simultaneously, so that both ends of the line can talk at the same time. However, carrier terminals are partly so designed that simultaneous seizure of both directions is not possible.

Of the previously mentioned message communication media, the telephone and the teleprinter are obviously best suited for communications of a general nature. The continuing expansion of the power supply systems brought about a commensurate increase in the requirement for such communication media as are intended for the supervision of technical operations in the power system. At the same time new communication circuits had to be provided for administrative purposes. In particular the circuits between the administration buildings of large companies may at times be overloaded to a point where certain highusage lines place the system operator in a position not unlike to that of a subscriber of the public network: the line is busy all the time. In a company-owned communication network, such a situation can be remedied by assigning entering privileges so that engineering calls are given priority over calls of an administrative nature.

In teleprinter traffic the emphasis is mostly not on bothway conversation between the two parties but, at least initially, on transmitting a message to the distant station. Two different areas of application evolved in the course of time, which determine the choice of the circuit to be used for a teleprinter connection. Where traffic is exchanged between agencies transmitting chiefly administrative messages, it is safe to assume that the communicating stations are located in places adequately served by networks of the postal authorities or of private communication companies. The public teleprinter network operazes exclusively on a sub-scriber-to-subscriber dialing basis. Traffic density is not nearly as high as in the telephone network. Moreover, a delay in the transmission of administrative messages, if the called party happens to be busy, is normally not felt to be too great an inconvenience. Since the messages are mostly destined for recipients outside the own company, such as government agencies, municipal administrations, and suppliers, this type of traffic is handled over the public teleprinter network.

If, on the other hand, a teleprinter connection must be established between vital points in the power network, as a supplement to the service telephone and exclusively for transmitting operational information, the construction of a private teleprinter system is fully justified. The corresponding parties are here the operators in the power and transformer stations who transmit either switching instructions and other short but important information, sent by teleprinter to avoid hearing errors, or long communications to relieve the telephone network. The idea of installing company-owned teleprinter circuits over the existing power lines which are independent of outside agencies and at any time available for service suggests itself immediately.

The load dispatchers in the control center of a large network have to react to a great many telemetering data. It appears to be reasonable to transfer this work load to a process control computer which accepts and processes this information. Whether the computer is to be engaged in the system management proper depends largely on the extent to which system operation is to be automated. Up to now computers are used to provide records of switch positions and alarms, to check them for their importance and to supply an indication as to whether the trouble is located in the power station, on the transmission lines or in the compound network. Besides, limit values are supervised, counter readings picked up, meter as well as counter indications recorded, and the essential information received in the course of a day stored and finally printed out in the form of a daily record. In addition to these supervising and recording functions, the computer may be entrusted with control duties, such as the control of switches, of ground fault compensation coils, of generators etc. The purpose is an "optimization of system operations", i. e. optimal distribution of the load to the individual generators combined with a check of the reliability of operation. In this connection line losses must be frequently taken into account and loading conditions supervised. All nodal points in the network must always be fed over at least two lines and many interdependances between indications and control information must be duly considered.

Obviously there is a considerable volume of data to be handled by the computer and the data must be transmitted in both directions at a

·comparatively high bit rate to retain the short response times of the telemetering system.

In this connection it should be mentioned that telephone circuits make up the bulk of the carrier communication circuits operated over power lines, although the percentage of the telemetering circuits is continuously increasing. Telephony has been somewhat relegated to the background in this section of the book, as compared with carrier terminals for telemetry. This is merely due to the fact that the author wanted to bring into relief the essential differences which must be considered when selecting the appropriate equipment for a given task.

3. Communication Circuits in the Power Network

For the carrier communication service of electric power plants it is the power line proper which is used for transmitting the HF currents. In most cases it provides the shortest connection between the communicating stations. Compared with open-wire communication lines, it has a better insulation and an appreciably higher mechanical strength. Damage caused by adverse weather conditions is therefore a very rare occurrence. Besides, a power line carrier communication system saves the cost of constructing a separate line.

The HF currents must be impressed upon and tapped from the power lines without causing a hazard to the communication equipment or to the operating personnel, and without any appreciable loss in power transmission.

To transmit these HF communication currents to the desired point, special paths must be formed in the power line network (Fig. 2). "Wave traps" are installed at those points where the communication currents might be dissipated in switching stations or line sections outside the transmission path, and "coupling media" where communication equipment is to be connected to the power line, or where the communication channel is to by-pass a station of the power network.

"Spur lines" deviating from the message path are blocked to the communication current and stations located along the transmission route are by-passed. A "by-pass circuit" consists of wave traps which prevent undesirable losses in the incoming and in the outgoing HF energy, as would be caused by the bus-bars of the station to be bridged, and of interconnected coupling facilities, which are arranged ahead of the traps to form the desired by-pass for the HF currents.

Compared with an open-wire communication line, operation over power lines is complicated by the fact that the communication system must continue to work properly even when power transmission is interrupted and the power line is connected to ground. Dissipation of HF currents in the ground and the resulting excessive additional attenuation of the message signals are avoided by traps inserted in the ground wires. Where power lines can be grounded in the switching stations, these traps are already available, because they are initially provided for regular service. If overhead lines have to be grounded en route because of repair work, signal transmission over the grounded power line can be maintained by means of portable traps.

The traps and coupling media required for providing a carrier communication channel on a power line are collectively termed "line equipment". While the traps are to allow the power current to easily pass

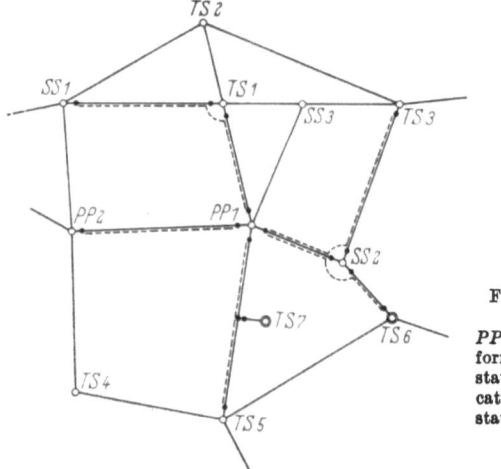

Fig. 2. Schematic layout of a power network.
PP Power plants; *TS* Transformer stations; *SS* Switching stations; ----- Carrier communication circuits; ○ Power network stations; ● Terminal points of carrier path (wave traps)

through them and to block the HF energy, the purpose of the coupling media consists in blocking the power current and allowing the communication current to pass. The operating frequencies of power networks range between $16^1/_2$ Hz and 60 Hz and those of the carrier communication system between 15 kHz and 500 kHz. Suitable blocking means are coils (inductances), rated for the operating current of the power plant, while capacitors, rated for the operating voltage of the power plant, are used for coupling purposes. Whereas the traps are self-contained units which are suspended from the power line on insulators or erected on pedestal-type insulators rated for the existing power voltage, the coupling media consist mostly of three constructional units: the "coupling capacitor", the "protective devices", and the "coupling filter" (also called "line tuning unit").

Attempts have been made in various countries to save the cost of the line equipment entirely and to use insulated ground wires along the route for carrier communication. The best solution, from a technical point of view, is obtained with lines having two ground wires. If these wires are steel wires, their surface must be coated with copper. However, for reasons connected with power engineering requirements ground wires are frequently steel-core aluminum wires, same as the phase conductors [29]. Arresters mounted en route on the turrets, which operate at voltages of about 10 kV, ensure that the protective effect of the ground wires for power current operation is not impaired. Ground faults are likely to affect carrier frequency operation. The degree to which such disturbances can be tolerated depends on the nature of the messages to be transmitted. Finally, there remains the question up to what distance the saving of coupling capacitors and wave traps means a real economy compared with the cost of an insulated ground wire with voltage limiters which is higher than that of a normal ground wire.

3.1 Coupling Circuits

During the early stages of power line carrier transmission in 1920, capacitors of sufficient rating, designed to withstand high voltages, were not yet available. Capacitive coupling was at that time accomplished with the aid of antennas, similar to those as were used in radio transmission. These antennas were mounted below the power lines on the tower structures. Antennas used for "antenna coupling" exceeded 300 feet in length and had to be tuned to the carrier frequencies employed. Since the capacitance of the switching plant enters into the tuning result, any changes in the switching condition brought about undesirable detuning.

Antenna coupling is rather inefficient and the entire set-up is easily affected by radiation from radio transmitters. Today, its use is therefore restricted to mobile communication equipment employed by line crews to establish a connection between their construction site and the nearest fixed-plant carrier telephone station, or for train-to-wayside communication with the wire lines along the railroad track used for coupling.

Efforts were made to take advantage of existing power plant facilities whenever possible, to avoid the necessity of special coupling media with their inherent error sources, and to save money. Lightning discharge wires, grounded by way of HF traps up to the second or third tower, proved at the beginning of this development to be just as inefficient as antenna coupling. Besides, these discharge wires were practically always made of steel and this led to a further loss of carrier energy. The capacitance of the insulated bushings proved to be insufficient for coupling carrier communication equipment to the power lines. Of all the methods tried out at that time only one has come to be definitely accepted: the use of a power capacitor both for coupling and measuring purposes, i. e. the capacitive voltage transformer with connections for carrier communication equipment [35].

Attempts were made right from the outset to use also power capacitors, aside from coupling antennas. This was chiefly due to the fact that some of the agencies engaged in this development refused to regard this method of signal transmission as belonging to the field of radio engineering. Instead, they classified it as carrier communication over physical lines. Hard paper capacitors were developed, which could be used only indoors. Porcelain capacitors for outdoor installation turned out to be too expensive at higher operating voltages. Since 1930, oil-paper capacitors have been used exclusively for line coupling in fixed-plant installations. From the early times of development stemmed the idea that the cost of a capacitor was determined chiefly by its capacitance rating. This is the reason why capacitors with not more than 550 pF and 1100 pF were built at that time. Today a rating of 2200 pF has been generally accepted as a minimum. Coupling capacitors are now availabel for up to 4400 pF, and capacitive voltage transformers may have still higher ratings (up to approx. 11 000 pF). Only for even higher ratings the limits are fixed by economy considerations and also by the fact that the leakage currents tend to increase with the capacitance rating.

In most countries standards have been established for ratings and tests of power capacitors which must be met by the coupling capacitors. International recommendations have also been worked out [1]. As far as the capacitive voltage transformers are concerned, they must also meet the standards set up for measuring transformers. All of these specifications are based on a power frequency of 50 Hz or 60 Hz. Guiding rules have been laid down in some countries to cover characteristics in the carrier frequency range. An international agreement is prepared by the IEC.

In outdoor switchyards capacitors for low operating voltages are suspension-mounted, while pedestal-mounted versions are preferred for higher voltages (Fig. 3). Some coupling capacitors are designed for mounting wave traps on their top (Fig. 4). Indoor capacitors, suitable only for low operating voltages, are increasingly falling into disuse and are now only occasionally manufactured. Today the preferred method is outdoor coupling at the mast in front of the station.

It is normal practice to give the value of the phase-to-phase voltage when stating the voltage of a power line. A 110 kV line, for instance, has a phase voltage of $110 : \sqrt{3} = 63.5$ kV with a wye-connected transformer. In the event of a ground fault on one conductor, the voltage against ground in the two remaining conductors may come up to a value which approximates the interconnected voltage, provided the star point is grounded by way of Petersen coils. Under normal operating conditions, a coupling capacitor is required to withstand merely the phase voltage. Anticipating the possibility of a ground fault, it must be rated for the interconnected voltage, i. e. for the full nominal voltage of the power line.

Power transmission networks have the neutral solidly grounded when they carry voltages higher than 150 kV. A coupling capacitor interposed between one conductor and ground may then be rated for a voltage lower

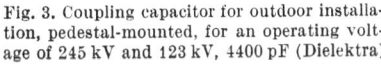

Fig. 3. Coupling capacitor for outdoor installation, pedestal-mounted, for an operating voltage of 245 kV and 123 kV, 4400 pF (Dielektra)

Fig. 4. Coupling capacitor for outdoor installation, pedestal-mounted with wave trap on top, for an operating voltage of 110 kV, 4400 pF, operating current 400 amps (Siemens A.G.)

than the nominal power line voltage. In accordance with the rating principles established for power line equipment, a value of 0.8 times the interconnected voltage is used for calculations.

Carrier equipment was originally connected to the line by means of a coupling capacitor tuned to one or two frequencies only and, as a result of this, it was not possible to connect more than a single carrier terminal.

Since 1935 however a filter network has been generally used instead
of the resonant circuit. The coupling capacitor thereby forms part of a
filter which allows a wide frequency band to pass (Fig. 5). In this way
a number of carrier terminals can be connected in parallel to the coupling
unit with a resultant saving in cost. However, the assignment of carrier

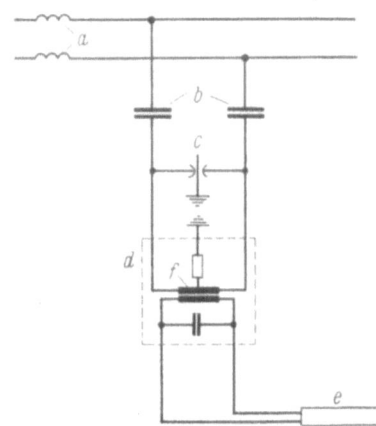

Fig. 5. Coupling circuit with bandpass
filter (Siemens A. G.)

a Wave traps; b Coupling capacitors;
c Coarse voltage arrester; d Coupling
filter; e Entry cable; f Insulating transformer rated for 5000 volts, primary to
secondary

frequencies in a power network may involve the necessity of using directional filters to limit the passband range to certain frequency bands
(see page 42).

From what has been said above one can see that line coupling units
have a twofold task to perform:

a) They are to protect the communication system from dangerous
overvoltages.

b) They are to allow a wide carrier frequency band to pass with a
minimum of loss from the power line to the communication equipment.

Both these problems and their solution can best be described in every
detail with the aid of a circuit schematic (Fig. 6), which is under discussion in Study Committee No. 35 of the CIGRE, in preparation of an
international recommendation for the design of such coupling circuits.
The charging current of the coupling capacitor increases with the operating voltage and with the coupling capacitance (Fig. 7). It must be safely
conducted to ground by way of a drain coil. If the line at the grounding
side of the capacitor is interrupted, overvoltages up to the magnitude of
the operating voltage may occur on the equipment side of the capacitor.
Great stress is therefore placed on optimum reliability and a copper
conductor of at least $1/4''$ diameter is used in connection with 10 kV
supports, although this conductor carries normally only a small current
and no voltage against ground. The coarse voltage arrester (air gap)

rated for an arc-over voltage of approx. 2 kV is provided to initiate, in the case of travelling waves or an interruption of the ground conductor, immediate grounding at the point where the overvoltage occurs. A grounding switch permits the entire coupling circuit to be grounded behind the coupling capacitor without interruption of power transmission

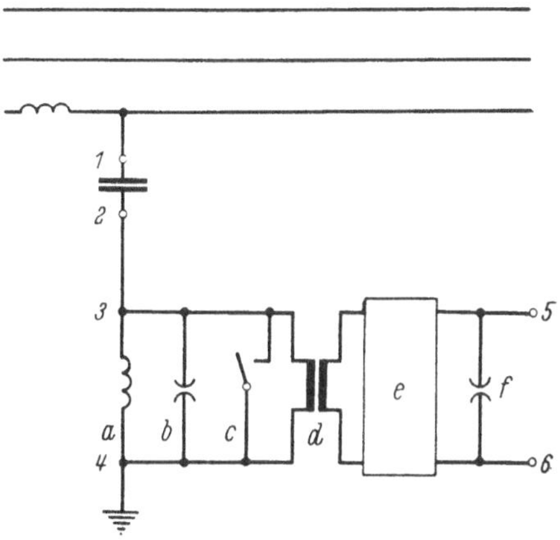

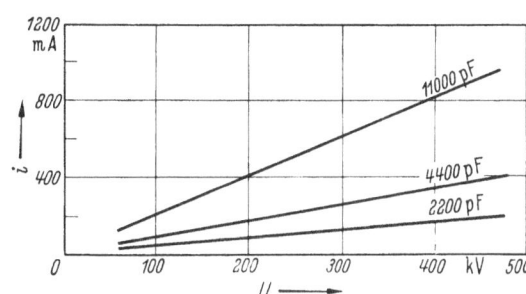

Fig. 6. Configuration of a coupling circuit with bandpass filter.

a Drain coil; b Coarse voltage arrester; c Grounding switch; d Insulating transformer; e Tuning elements; f Voltage arrester; 1 H T terminals of coupling capacitor; 2 Low tension terminals of coupling capacitor; 3, 4 Line terminals of coupling filter; 5, 6 Cable terminals of coupling filter

if a trouble must be cleared in the communication equipment. The insulating transformer is rated for a dielectric strength of 7 kV between primary and secondary. The tuning elements are to match the characteristic impedance of the power line to that of the coupling circuit for as large a portion of the carrier frequency range as possible. A further voltage arrester may be interposed between the outdoor coupling facilities and the entry cable which connects the carrier communication terminal in the station building.

Fig. 7. Charging current of coupling capacitor as a function of operating voltage and capacitance, power frequency 50 Hz

This entry cable has a dielectric strength of about 2 kV against ground. For distances up to several hundred yards a low-capacitance, single-pair telephone cable will do in most cases. Only where large distances have to be spanned will it be necessary to provide special HF cable or open-wire lines, to avoid undue attenuation of the HF currents. This applies in the case of distances of only a few hundred yards if high-carrier frequencies (400 kHz to 500 kHz), are transmitted. In cases where the entry cable is run over a large distance in parallel with power cables in one trench, or if open-wire leads cross a switching plant, further

a

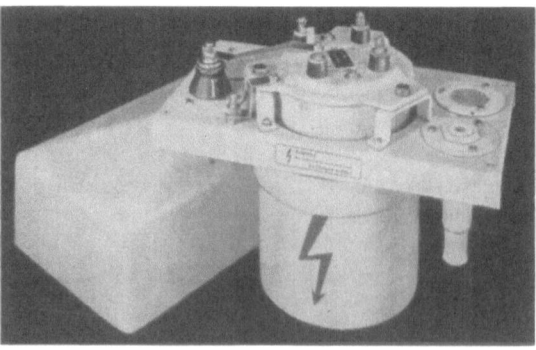

b

Figs. 8a and 8b. Coupling filter for outdoor installation (Siemens A.G.)
a) With cover; b) Cover removed

protection must be provided immediately before entry into the carrier terminal to prevent electric shock hazard.

Cables are partly of the single-conductor type, the wire serving for carrier energy transmission and the cable sheath as return path. Two-pole cables, i. e. cables containing a wire pair, are also used in which case the output and input circuits between coupling media and carrier terminal can be of the ground-balanced type, so that overvoltages, as might be produced by parallelling power lines inside the switching plant or by compensating currents, will have no influence.

The safety of the operating personnel and of the communication equipment depends essentially on the coupling circuit. Besides, its influence on the quality of message transmission must not be neglected. Much effort has therefore been expended on perfecting the design of the coupling circuits. On the other hand, equipment complication and costs

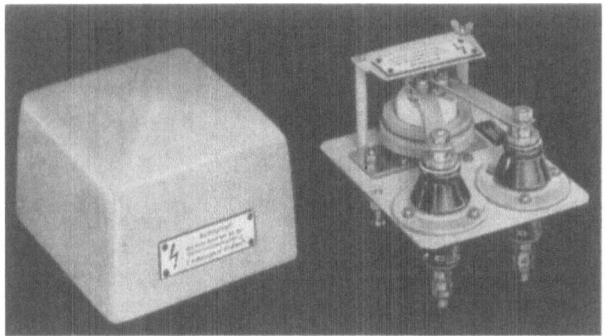

Fig. 9. Coarse voltage arrester for outdoor installation (Siemens A.G.)

Fig. 10. Coupling protector (Siemens A.G.)

rise with each attempt to consider all possible technical angles. Any solution to this problem must be a compromise between technical requirements and endeavors to keep the costs down. Thus, for instance, a minimum of equipment complexity is achieved with the previously mentioned circuit configuration (Fig. 5), where the functions of the drain coil, of the insulating transformer, and of the matching transformer are consolidated in one single unit. The grounding switch has been entirely omitted. It must be additionally installed or replaced by a ground wire in case of need. The question whether preference should be given to a maximum of simplicity as in this circuit which, having only a few components, is less susceptible to trouble (Figs. 8 and 9), or to a more elaborate arrangement (Figs. 8 and 10), is decided by the requirements to be

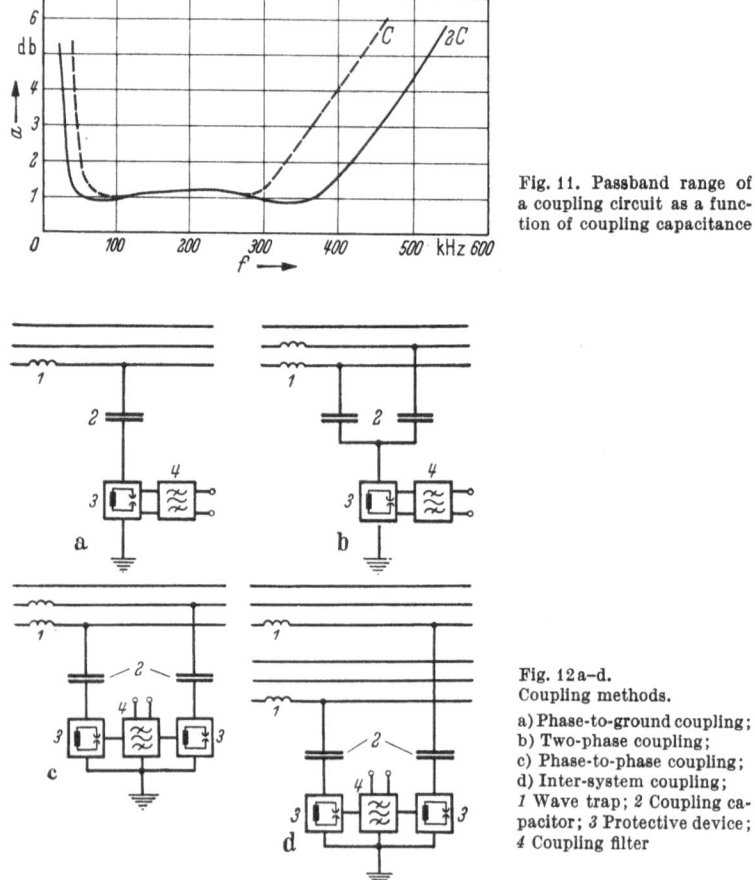

Fig. 11. Passband range of a coupling circuit as a function of coupling capacitance

Fig. 12 a–d.
Coupling methods.

a) Phase-to-ground coupling;
b) Two-phase coupling;
c) Phase-to-phase coupling;
d) Inter-system coupling;
1 Wave trap; 2 Coupling capacitor; 3 Protective device; 4 Coupling filter

met or by the applicable safety standards. A uniform concept for al countries is still a long way from acceptance.

The passband range of a filter circuit for these purposes depends, aside from the circuit arrangement, only on the magnitude of the coupling capacitance (Fig. 11). The larger the capacitance, the lower are the frequencies which can be transmitted. An increase in capacitance involves an increase in the charging current. This, in turn, will entail the necessity of using a separate drain coil in the coupling circuit, because a high insulating rating of the transformer and its tuning by means of the matching elements cannot be readily reconciled with the task of discharging appreciable currents over a prolonged period. If the coil is inserted in the ground lead of a capacitive voltage transformer, it will be incorporated in the latter, partly to comply with grounding regulations for HT equipment, and partly to keep its influence on the measuring accuracy more readily within the permissible limits (see page 26).

Carrier communication terminals may be connected to two conductors of a power line. Depending on the circuitry (Fig. 12), a distinction is made between "two-phase coupling", formerly adopted to improve reliability of transmission in the event of a breakage of one of the coupled conductors, "phase-to-phase coupling", using again two conductors of a three-phase system, and "inter-system coupling", where one conductor each of two three-phase systems suspended from a common tower is used. A connection may also be made between a single phase of a power line and ground. This method is called "phase-to-ground coupling". Compared with the other coupling methods, only half the number of coupling capacitors and wave traps is here required.

With phase-to-phase coupling (and inter-system coupling), one may speak of metallic go and return lines for the carrier currents. The third conductor of the three-phase system, which contains no wave traps for blocking the communication currents (or the four untrapped conductors of two parallelling three-phase systems), have no appreciable influence on communication current transmission. As a result, the switching conditions of power stations will not affect carrier communication to a degree worth mentioning.

From voice-frequency telephony we know that speech can be transmitted over a single wire with ground return. When carrier communication terminals were connected to only one conductor, it was originally believed that the ground would substitute as return leg for the carrier currents. Theoretical deliberations and practical experiments proved, however, that carrier communication with phase-to-ground coupling is actually accomplished in another way. If only a single conductor and ground were engaged in transmission, the message would not proceed very far because of eddy current losses in the ground. In a three-phase system, however, the two non-coupled conductors take also part in transmission, whereby the distance which can be spanned with phase-to-ground coupling is increased (see page 48).

While two-phase coupling also provides for coupling to two conductors, the two coupling capacitors are, however, connected in parallel to the same coupling filter. This arrangement ensures a higher measure of reliability than phase-to-ground coupling in the event one of the two coupled conductors is damaged. The assumption that connecting two conductors in parallel (having in mind the doubling of the cross section in voice frequency transmission) would improve transmission conditions is wrong.

To answer the question as to which type of coupling should be adopted in a specific case, the characteristics of the various coupling methods must be remembered:

		Equipment requirements	Attenuation	Reliability in case of broken conductor	Monitoring possibility, interference by radio transmissions
a	Phase-to-ground coupling	Minimum	Larger than c and d	Minimum	Larger than c and d
b	Two-phase coupling	$2 \times a$	Larger than a	Larger than a	Larger than a
c	Phase-to-phase coupling	$2 \times a$	Minimum	As b	Minimum
d	Inter-system coupling	$2 \times a$	As c	As b	As c

Fig. 13. Characteristics of the different coupling methods

Two-phase coupling has been abandoned because of the excessive coupling loss. Phase-to-ground coupling is preferred for economical reasons. From a technical point of view it may be regarded as adequate if the distance to be covered is not too long and where no extra-high voltage lines with an inherently high noise level are involved. Breakage of the coupled conductor may result in a breakdown of the communication system if the faulted conductor is interrupted (and possibly grounded) near a coupling or a by-pass point. If the trouble occurred some distance away from the coupling points, the interruption of the coupled conductor may cause a major increase in attenuation, but carrier communication need not necessarily be rendered impossible.

Phase-to-phase coupling and inter-system coupling afford a higher measure of dependability in the event of a broken conductor, because signal transmission continues over the other coupled conductor. Inter-system coupling offers the additional advantage that one of the two three-phase systems can be grounded en route without interposition of wave traps and the carrier communication system will still remain operative. The benefit of inter-system coupling will get lost, however, if an attempt is made to couple the carrier terminals to three-phase systems strung on different supporting structures, because a wide separation of the coupled conductors is likely to produce annoying delay time variations. Moreover, in view of the different distribution of capacitances between conductors and ground, inter-system coupling would turn out to be in fact a twofold phase-to-ground coupling arrangement.

Because of the lower attenuation, phase-to-phase and inter-system coupling are adopted where large distances have to be spanned, or where a high interference level is encountered. Practical examples are all lines of 220 kV and higher. If a number of by-pass circuits have to be arranged along one line section or excessive sleet formation must be expected on power lines during the winter months, phase-to-phase coupling is again employed to avoid an intolerable overall attenuation of the transmission path under these adverse conditions.

Aside from signal attenuation, there might be other compelling reasons for phase-to-phase coupling. Considerations of economy are here passed over and the emphasis is shifted to maximum reliability in the event of a broken conductor. Operation of a line protection system with carrier channels is a typical example.

All connecting lines within the coupling arrangement should be as short as possible and of the open-wire type, to avoid an adverse influence on the characteristics of the bandpass filter in the carrier frequency range. With phase-to-ground coupling, this requirement can be easily met in practically all cases. With phase-to-phase and inter-system coupling, however, an undesirable arrangement may result in outdoor switchyards if the two coupled conductors are widely spaced, and a two-conductor coupling filter, common to both coupling capacitors, is to be located approximately in the center between the two coupling points. Under these circumstances it will be better to use a single-conductor coupling filter for each coupling capacitor, and to interconnect both filters by a cable. Obviously, configurations of this type prevail in large outdoor switchyards of 220 kV and higher operating voltages.

Capacitive voltage transformers are frequently used for coupling the carrier terminals to high-voltage power lines. While coupling capacitors are invariably connected immediately at the end of the line, i. e. ahead of the first disconnect switch, to maintain message communication also when the power line itself is disconnected, the necessity of this arrangement must be particularly stressed in the case capacitive voltage transformers are installed. If a capacitive voltage transformer, for reasons connected with power line operation or with measuring techniques, is inserted behind the disconnect switch, it can no longer serve its purpose as coupling medium for the carrier equipment, because message communication is interrupted when this switch is open.

A capacitive voltage transformer (Fig. 14) is essentially a voltage divider designed as a power capacitor with measuring tap. Magnitude and phase of the measuring voltage at the output terminals of the auxiliary transformer are largely made independent of the load with the aid of a compensating choke coil, which is resonated to the sum of the two partial capacitances. The choke coil must not constitute an excessive shunt (due to its capacitances) for the carrier communication currents and a HF choke coil must be connected ahead of it, if necessary.

For pure measuring purposes, the resultant capacitance must be the larger the lower the operating voltage to be measured is, and the higher the power demands on the transformer are at a given accuracy, or the higher the transformer accuracy demands are at a given rated power. In accuracy class 1 (corresponds to 1 % measuring accuracy within 80 % to

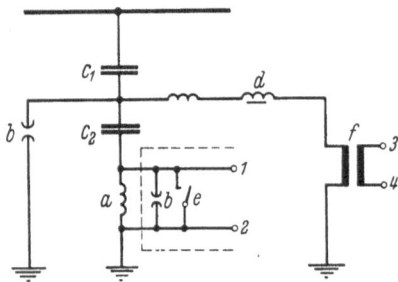

Fig. 14. Basic circuitry of a capacitive voltage transformer with connections for a carrier communication terminal.

a Drain coil; b Overvoltage arrester; C_1 Power capacitor; C_2 Measuring capacitor; d Compensating choke; e Grounding switch; f Auxiliary transformer; g Carrier choke; $1, 2$ to the line terminals of the coupling filter; $3, 4$ to the measuring circuits

120 % of rated voltage) a capacitance of 4400 pF will therefore be generally available at the input terminals of the coupling filter if operating voltage and rated power are 110 kV and 120 VA, respectively, just to give an example. With an operating voltage of 220 kV, but otherwise equal conditions, a capacitance of only 2200 pF will do.

As can be seen, the capacitance required for measuring purposes is of the same order of magnitude as is required for line coupling. When selecting the capacitance value, it is therefore important for the sake of the most economical overall solution to find a common denominator for measuring duties and carrier frequency coupling. Constructionally, there are two options for designing a capacitive voltage transformer. The first provides for accommodation of the two capacitors within a common porcelain body ensuring uniform dielectric conditions. Temperature variations will then affect the two capacitors to an equal degree, so that the division ratio is maintained and an additional measuring error avoided. This option may be chosen whenever the measuring and the communication facilities are installed at the same time. Adoption of the second option is advisable if line coupling is performed a considerable time before the measuring problem has to be solved and, to mention another viewpoint, if the investment costs are to be defrayed in two stages. In this case coupling capacitor c_1 is purchased first and only at a later stage the measuring attachment comprising capacitor c_2, which is constructionally combined with the auxiliary transformer f and the compensating and carrier chokes so as to form a single unit. As previously mentioned, this solution involves the danger of additional measuring errors, which are due to different effects of temperature variations in the two sections and the resultant disturbance of the division ratio $c_1 : c_2$.

The profit derived from using capacitive voltage transformers simultaneously for measuring purposes and for carrier coupling consists essentially in the saving of the coupling capacitors. Since one capacitive voltage transformer must be connected to each conductor of a power system for measuring purposes, the possibility of coupling the carrier communication terminal to two conductors will always exist. This "built-in" benefit of phase-to-phase or inter-system coupling is therefore available at reduced cost. The advantages are substantially increased dependability in the case of wire breakage, lower attenuation, and greater immunity to interference from radio transmitters (see page 24). In spite of this, lack of funds compelled system operators in many cases to decide in favor of phase-to-ground coupling, since only half the number of coupling capacitors and wave traps is required. With capacitive voltage transformers engaged for line coupling, the monetary pressure is greatly relieved.

Capacitive voltage transformers, integrated with current transformers to form a single unit, may also be used as combined transformers for coupling the carrier equipment. The primary of the current transformer is here permanently connected at one side with the HT terminal of the voltage transformer through which the carrier frequency enters. Since all attempts to design current transformers (with iron core) so as to make them function as a trap for the communication currents have been unsuccessful so far, a trap must be provided behind the combined transformer and ahead of the entry into the switching plant. Furthermore, it will be necessary to verify that the inherent capacitance of the current transformer section does not involve an excessive loss in carrier energy.

Coupling capacitors and capacitive voltage transformers have an inherent inductance which, in the case of lower carrier frequencies, may be regarded as being in series with the capacitance in an equivalent circuit diagram. It effects an increase in the coupling capacitance. For high frequencies the effect is that of a parallel resonance. To avoid a blocking effect of the coupling circuit, i. e. the contrary of what the cpouling circuit is supposed to do, it is necessary that this "resonance frequency of the coupling capacitor" is situated high enough above the transmission range, i. e. at a frequency about 1.5 to 2 times the highest frequency to be transmitted. In the course of IEC discussions aiming at an internatio nal convention a frequency of 800 kHz has been suggested to permit an economical design of capacitors rated for operating voltages of more than 220 kV. At lower operating voltages a resonance frequency of about 1 MHz has been attained.

3.2 Wave Traps

The power network station at the end of the communication path is connected in parallel with the coupling circuit. The impedance which it presents to the carrier frequencies depends on its constructional layout and on its switching condition. Major capacitances of the bus bars entail a considerable loss while transformers, under certain circumstances, are free from this disadvantage. Attempts have therefore been made during the earliest stages of development to go without special carrier frequency traps. It did not take much time, however, to realize that the impedance of a station is a rather uncertain factor in carrier communication and that it is safer to provide special means to prevent the impedance from falling below a certain minimum. The solution found after some detours [23] consisted in arranging choke coils ahead of the entry to a station, or to a network section not belonging to the message path. Short tie lines, the socalled spurs, are particularly annoying because of resonance phenomena seriously impairing carrier communication (Appendix 9.1).

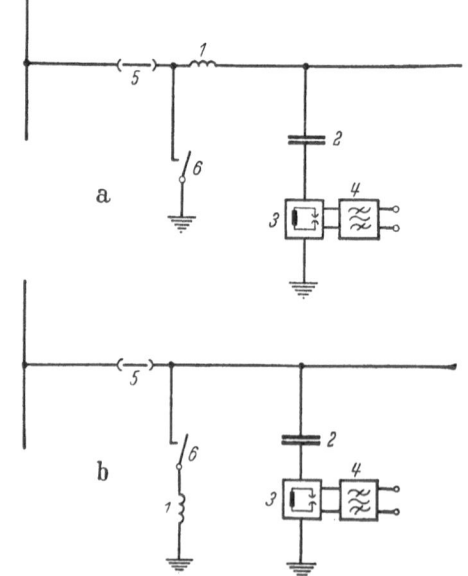

Fig. 15a and b. Arrangement of wave traps in a station.

a) Wave trap in the line carrying the operating current; b) Wave trap in the lead to the grounding switch; 1 Wave trap; 2 Coupling capacitor; 3 Protective device; 4 Coupling filter; 5 Disconnect switch; 6 Grounding switch

A wave trap is connected directly to the power line and must therefore have an insulation resistance rated for the power voltage. Above all, it must be rated for the full operating current. In the event of power system faults it should display the same short circuit resistance as the power system, i. e. it should be able to withstand the same dynamic and thermal limit current as the other power plant equipment. Size and cost of wave traps are determined by these requirements. Various attempts have therfore been made, with the ultimate purpose of cost reduction, to arrange the wave traps inside the station rather than at its entry in an effort to arrive at more favorable conditions for their rating.

Assuming that power transformers may have an impedance sufficiently large to permit blocking [27] and that the total impedance of the station is not materially changed by switching operations, it is possible to use a small trap connected to the ground line and rated only for a low current (Fig. 15). Operation of the communication system would here be ensured also in the case the line is grounded in the station. But this solution cannot be generally adopted because the stations frequently have too low an impedance, or because their impedance is greatly affected by switching operations. Besides, regulations in many countries preclude the insertion of "foreign" elements in the ground wire. Another disadvantage of this arrangement is the fact that the carrier signals continue on their way and flow into other line sections. This is the reason why wave traps are in most cases inserted between the coupling capacitor and the disconnect switch.

Wave traps rated for the operating current used to be manufactured in a rather wide variety of versions by the individual manufacturers. To achieve a certain measure of clarity and uniformity, efforts are made to establish universally applicable specifications covering ratings and testing procedures, as in the case of the coupling circuits. The proposals intended to establish standards for wave traps are based on the applicable electric power standards.

The IEC recommends a certain grading as to rated currents and associated short circuit resistances (Fig. 16). Rated short-duration current

| | Minimum values of limit current | | | |
| | Series 1 | | Series 2 | |
Rated current A	Rated short-duration current kA rms	First current amplitude kA	Rated short-duration current kA rms	First current amplitude kA
100	2.5	6	3	13
200	5	13	10	26
400	10	26	16	41
630	16	41	20	51
800	20	51	25	64
1000	25	64	31.5	81
1250	31.5	81	40	102
1600	40	102	50	102
2000	40	102	50	102
4000	50	118	63	162

Inductance: $0.2 - 0.25 - 0.4 - 0.5 - 1.0 - 2.0$ mH

Fig. 16. IEC-Recommendations for wave trap rating (cursive figures: preferred values)

refers to the maximum rms value of the current stated in kA, whose thermal effect the wave trap will withstand for a period of one second, following a continuous loading with rated current under standardized temperature conditions. Besides, the wave trap must be able to withstand the first current amplitude which is 2.55-times the rated short-duration current.

Some manufacturers produced traps designed for current ratings lower than those indicated in the table as well as traps for many intermediate values. Traps for still higher ratings have also been manufactured in recent years. While a certain overload carrying capacity in continuous operation was initially required, similar to the overload rating of current transformers, this practice has been abandoned during the last years. In the case of the current transformer, the stated overload carrying capacity applies to a certain measuring accuracy class, so that an indication of this rating is well justified. Wave traps, on the other hand, have in any case to be rated for the maximum load. In the past, wave traps used to be rated for lower limit currents.

Fig. 17. Resonant trap for an operating current of 400 amps, 0.2 mH (Siemens A.G.)

Wave traps of recent design and intended for high nominal currents are now required to withstand the ever increasing short-duration currents. This tendency is in line with the development in electric power engineering. The short-circuit power in the power networks increases steadily. A carrier frequency trap may be installed at the input to a large power station. It is possible that a ground fault or a short-circuit fault occurs in the line only a short distance away from the trap. The trap is then subjected to a current surge which corresponds practically to the full generator output.

The carrier frequency characteristic of wave traps is determined by the inductance of the traps. A distinction is made between "resonant

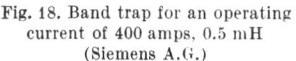

Fig. 18. Band trap for an operating current of 400 amps, 0.5 mH (Siemens A.G.)

Fig. 19. Band trap for an operating current of 2500 amps, 4.0 mH (Allgemeine Elektricitäts-Gesellschaft)

traps", which block only one or two carrier frequency channels (inductance approximately 0.2 mH) and "band traps" which block the available carrier frequency range completely, or a large portion thereof (inductance up to 2.0 mH).

Disregarding the rated short-duration current, the amount of conductor material required, i. e. the larger part of the cost of a wave trap, is determined by the current rating and the inductance, while the mechanical design and the cost thereby involved depend on the peak value of the first amplitude of the rated short-duration current. Whereas resonant traps have a single-layer cylindrical coil (Fig. 17), the band traps with their ten times higher inductance have a disc shape, similar to that of drain coils (Figs. 18, 19). The conductor material is copper-profiles or overhead power wires. Many other forms of traps not described in this book have been developed in the course of time.

The earlier-type traps were the resonant traps. Until 1940 this was the only type in general use, as the number of carrier frequency bands required at that time was very small. In case of need, two resonant traps were arranged one below the other. A rather awkward mounting arrangement resulted where three resonant traps had to be connected in series.

The coils were at first resonated to one frequency only (Fig. 20a). To reduce the size of the power current carrying coil and of the tuning capacitor to a minimum, a secondary circuit, tuned to the carrier frequency and containing a coil with a higher number of turns, was inductively coupled with a coil having a smaller number of turns which was connected to the line (Fig. 20b). In the event of traveling waves, the voltage induced in the secondary coil was, however, so high that the capacitors with a solid dielectric material, connected in parallel with the variable capacitor to achieve sufficient capacitance, were frequently punctured. This method of tuning was therefore abandoned very soon.

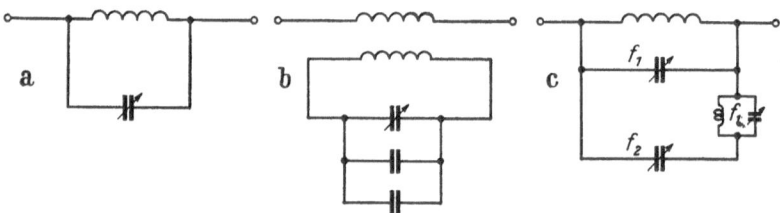

Fig. 20a–c. Examples of resonant trap circuits.
a) Single-frequency resonant trap; b) Single-frequency resonant trap with tuned secondary; c) Two-frequency resonant trap with parallel circuit tuning

Since the large majority of carrier communication systems is intended for speech transmission, this type of trap was soon replaced by a version tuned to two frequencies.

A frequently used circuit arrangement provides for an inductance of approximately 0.2 mH, to be initially resonated by means of a parallel capacitor to a frequency f_1 (Fig. 20c). A further capacitor, in series with a parallel-resonant circuit tuned to f_1, is also connected in parallel with the inductance and rated so that the entire circuit arrangement will block also the frequency f_2. Frequency f_1 is the higher, f_2 the lower of the two carrier frequencies.

With resonant tuning, the impedance for blocking the two carrier frequencies can be raised to a multiple of the characteristic impedance of the power line. This ensures uninterrupted carrier communication when the power line is grounded behind the traps, or when the switching condition of the line plant is changed.

In some wave traps the tuning media are accommodated inside the cylindrical coil. Other designs provide for a housing protected against rain to be installed close to the trap. Guard shields were formerly used to protect the entire wave trap against rain. These shields have been omitted for the sake of improved ventilation, when it became possible to make the surface of the coils resistant to adverse weather conditions.

Stringent requirements are imposed on the tuning capacitors. First of all they must be designed to withstand very high test voltages, because a high voltage drop occurs across the coil in the event of short circuit currents, and traveling waves might damage the trap. Defects on the tuning capacitors can be detected only through measurements in the carrier communication system, since the traps are under load during operation and therefore not accessible for direct measurements (Fig. 21). To replace a defective tuning capacitor, the power system must be deactivated and the wave trap removed and retuned. To avoid this inconvenience, tuning capacitors are rated for as high a voltage as is compatible with their dimensions and cost.

Fig. 21. Phase-to-phase broad-band coupling to a 750 kV line, wave traps for an operating current of 3150 amps (Brown, Boveri & Cie.)

The temperature stability of the tuning capacitors is another important factor. The considerable temperature changes to which the capacitors are exposed in outdoor switchyards must not cause any appreciable change in the capacitance rating to avoid undue detuning, i. e. a departure from the carrier frequency to be blocked.

Voltage arresters, which may be vacuum arresters, are inserted between the terminals of the coil to protect the capacitors. The arc-over voltage of these arresters lies, on the one hand, below the voltage for which tuning capacitors are rated and, on the other hand, above the voltage produced across the coil in the event of a short-circuit current surge. The tuning capacitors are thereby protected against the momentary overvoltages entailed by traveling waves. Sustained overvoltages, resulting from a short circuit surge, are not sufficiently high to cause the vacuum arrester to arc-over. This eliminates the danger of a destruction of the arrester.

In large switchyards the voltage arresters protecting the power equipment are arranged in the immediate vicinity of this equipment, mostly close to the power transformers. The wave traps for the carrier system, however, are located a considerable distance away, i. e. at the entry to the switchyard. To prevent the small arresters of the wave traps from being overloaded and destroyed by the large arresters of the power

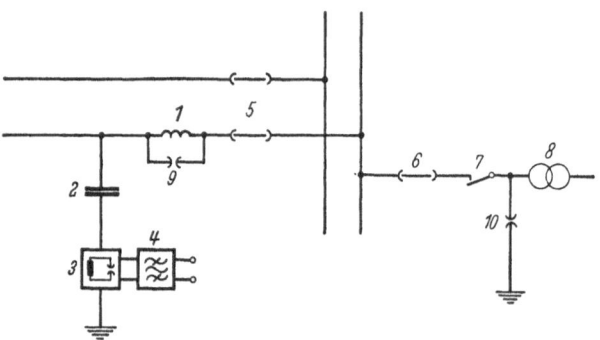

Fig. 22. Arrangement of overvoltage arrester in a station.

1 Wave trap; *2* Coupling capacitor; *3* Protective device; *4* Coupling filter; *5* Line disconnect switch; *6* Transformer disconnect switch; *7* Circuit breaker of transformer; *8* Transformer; *9* Overvoltage arrester of wave trap; *10* Overvoltage arrester of transformer

equipment arcing over, both types of arresters must be designed for the same discharge capacity (approx. 10 kA), although they are rated for greatly different arc-over voltages (Fig. 22).

Resonant traps are parallel-resonant circuits, consisting essentially of a coil and a shunt capacitor. An excessive difference in the impedance presented to the frequencies within the blocking range is highly undesirable, as it would impair the quality of transmission. An additional resistance is therefore provided to dampen the resonant circuit. The development of resonant traps has been carried on in an effort to block not only two, but as many carrier frequencies as possible. The desirability of using a comparatively expensive component, rated for the full operating current, for as many communication circuits as possible, initiated the development of coupling circuits including bandpass filters. The final result of these development efforts is the coupling capacitor. The passband range has here been extended by increasing the coupling capacitance.

The width of the blocking range increases with the inductance of the wave trap coil through which the power current flows. Tuning to individual carrier frequencies is not necessary and a few wave traps, tuned to different frequency bands, suffice to cover the entire frequency range assigned to carrier communication. In the event of a short circuit in the

power network these wave traps, having a higher inductance, will be
subjected to greater mechanical forces, and higher voltages will occur
across their tuning capacitors.

Coils rated for 2.0 mH have been installed to eliminate, where possible,
the need for tuning capacitors, so that operators must no longer be afraid
of the possibility of tuning capacitors being ruined by overvoltages and
inclement weather conditions. Commensurate with their inductance,
these coils must be very sturdy so as to withstand a short circuit current
surge. Depending on the inductance of the choke coil, its reactance will
suffice to cover a large portion of the carrier frequency range (Fig. 23).
Where national regulations permit only a part of this range to be used
(see page 51), a smaller inductance, say 1.0 mH, will also do. A blocking
impedance may be regarded as being adequate if it equals approximately
the characteristic impedance of the power line. At a frequency of 50 kHz,
for instance, a 2 mH wave trap will present an impedance of $\omega L = 628\ \Omega$.
If the lower limit of the available frequency range is as high as 100 kHz,
the same impedance can be attained with an inductance of not more than
1.0 mH.

If coils without capacitors are used as wave traps, the possibility of
trouble introduced by tuning media is eliminated. Such coils may, how-

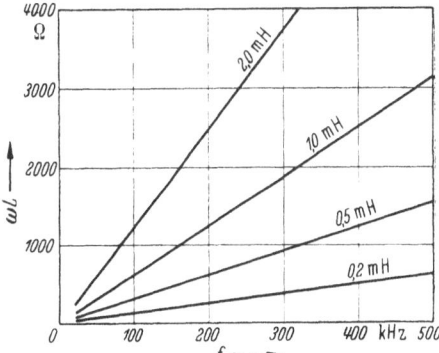

Fig. 23. Reactance of a
choke coil as a function
of carrier frequency

ever, form a series-resonant circuit with the capacitance of the bus bars
so that certain carrier frequencies will be allowed to pass. Practical
experience has shown, however, that this is a very rare occurrence with
an inductance of 2.0 mH and that it does not pay to provide each trap, as
a matter of principle, with means to counter such a possibility. It will
suffice to connect an attachment, consisting essentially of a damping
resistor and a low-load capacitor. With inductances of 1.0 mH or 0.5 mH
this risk is greater, so that these traps should always be provided with
tuning media.

An inductance of 2.0 mH becomes increasingly uneconomical the higher the operating currents and, above all, the higher the rated short-duration currents are. With a nominal current of 1600 A and a short-duration current of 100 kA, for instance, the cost will increase to five times the cost of a wave trap of the same inductance, but rated for a nominal current of 400 A and a short circuit current of 50 kA. Power line carrier operation would no longer be economically justified if pure inductances of 2.0 mH were to be used as wave traps. Where short-duration currents of 100 kA must be expected, smaller inductances are therefore used together with tuning media to obtain broadband traps. To save costs, many countries have made it a principle to retain this arrangement also for smaller short-duration currents. Complete freedom afforded by the 2.0 mH wave traps in the selection of carrier frequencies has, however, been sacrificed. Any rearrangement in the carrier communication system will involve modifications on the wave traps, a nuisance avoided with 2.0 mH wave traps.

The efficiency of a wave trap is assessed by its impedance or, to be more precise, by the shunt attenuation which it presents to the carrier communication system. The shunt attenuation depends not only on the design of the wave trap, but also on the impedance of the switching station (or on the trap coupling point) (Appendix 9.2).

It is not possible to design a coil completely free from capacitance. The inherent capacitance of a coil may be regarded as being connected in parallel with the inductance in an equivalent circuit diagram. Its magnitude is about 100 pF. In coils without tuning media (2.0 mH) it results in the "natural resonance frequency of the trap", mostly in the upper portion of the transmission range, between 250 kHz and 500 kHz, dependent on the design. For this point in the frequency range, the blocking resistance of an unresonated coil increases to a high value. It follows that the inherent capacitance serves the purpose of the trap. In the case of resonated traps (inductances 0.2 mH or 0.5 mH) the capacitances of the tuning circuits for high frequencies are somewhere near 1000 pF. Thus, the inherent capacitance of the coil is about 10 % of the tuning capacitance and must therefore be taken into account when resonating the traps.

A low-inductance choke coil of the type required for resonant traps is invariably designed as a single-layer cylindrical coil. This simple and inexpensive design is retained also for larger inductances as far as this is compatible with dimensions and material requirements. For higher inductances, a disc-shaped coil turned out to be a more favorable solution. The weight of wave traps increases rapidly with inductance and nominal current so that the traps must be mounted on pedestals and can no longer be suspended from wires or cross arms. Although installation on special insulating supports, which must be rated for the operating voltage, is more expensive than suspension mounting, it is frequently more compatible with the layout of switchyard facilities. In the case of very high short-duration currents, however, suspension mounting is again given preference because it provides for better absorption of the mechanical shocks caused by short circuits, including those in the trap leads.

The wave traps discussed so far are all dimensioned for practical operating purposes. They are characterized by the fact that their blocking impedance need not be higher than is required to reduce undesirable dissemination of carrier energy so that proper operation of the carrier terminals is ensured irrespective of the switching condition of the power system and a possible grounding of the power line by way of the traps. For economical and technical reasons, practical wave traps cannot be designed so as to provide a blocking efficiency comparable to full decoupling. Besides, connecting high-grade traps only to those conductors which are actually coupled would be to no avail. Since the power conductors are strung in parallel over wide distances, an appreciable portion of the carrier energy is taken over by the non-coupled conductors of the power system, and carried beyond the trapped points into other sections of the power network. Frequency allocation in a power distribution network is greatly complicated by this phenomenon because a network constitutes a mesh, which is metallically interconnected in all nodal points by way of the station bus bars. In spite of the wave traps, a frequency band used in one section of the network cannot be used again in another section, unless the two sections are separated by a considerable distance.

Another type of trap, a type which offers a markedly increased blocking efficiency is therefore used [15]. Such "cross-talk traps" or "network decoupling circuits" are needed at certain points of an extensive interconnected power network to form separate HF sub-networks which are then effectively decoupled so that all frequencies of the entire carrier range can be used in each individual sub-network. In other words, cross-talk traps have the duty of subdividing a large intermeshed network, for which more carrier channels are required than can be accommodated in the range between 15 kHz and 500 kHz, into a number of smaller sub-networks for the purpose of carrier communication.

Decoupling may be accomplished in different ways. Power transformers, cable sections in the power line, line sections with iron sheathing or network decoupling circuits consisting of coils and capacitors may be employed. The expenses thereby incurred are so high that, where possible, only such means are used, as offer themselves in the form of integral parts of the power system, i. e. power transformers, and long power cables between tap-off points and sub-stations [26]. Of the media serving exclusively for network decoupling, power lines with iron sheath (Appendix 9.3) have as yet not been used.

A power transformer is a satisfactory device for decoupling. Transfer of communication currents from one side to the other is practically excluded. Power networks separated by means of transformers permit independent frequency allocation schemes to be drawn up for the various HF communication systems. Nevertheless, it is not advisable to use identical carrier frequencies on lines carrying different operating voltages to the same sub-station, because in many cases undesirable coupling of networks of different operating voltages cannot be fully avoided at the entry of the power lines in the sub-station. To reduce the short circuit power, transformers of a 1 : 1 ratio have been interposed between larger sections of a power network. Attempts to have these insulating transformers designed so that they would double as cross-talk traps for the HF communication currents and, over and above this, the demand for installing such power transformers solely to satisfy carrier communication requirements, failed to yield any substantial results so far because of the high costs involved.

A special version is the "portable wave traps", which are used where a disconnected power line has to be grounded en route. When a portable HF telephone equipment is set up to permit telephone conversation between the line crew and the nearest fixed-plant station (see page 106), portable wave traps can be connected to the ground lines, provided to ground the power line for reasons of safety at both sides of the construction site.

The operating conditions of portable traps are therefore different from those of permanently installed wave traps. Portable traps are not required to carry the power line current and, depending on their location within the power network, must be used for any of the various carrier frequency channels, i. e. they must be capable of blocking all frequencies of the available range. Moreover, when a trap has to be installed en route, doubts might arise as to which of the three conductors of the power system the carrier communication equipment has actually been coupled. It is therefore common practice to block all three phases against ground by means of traps inserted in the ground line. When repair work has to be carried out on a power line, the three conductors are normally shorted and connected to a common ground. A portable two-pole all-wave trap may then be inserted in the ground line. This arrangement will be satisfactory if a single carrier communication equipment is coupled to one power conductor only. In the majority of cases, however, carrier terminals are coupled to more than one power conductor. It may happen that one carrier terminal is coupled to two conductors, or that independent carrier systems are coupled to one power conductor each. To allow for all these cases, the portable all-wave trap is of the three-phase type.

To ensure that these wave traps are properly installed by the line crews, the traps must be easy to transport and means must be provided for convenient suspension from a tower structure and for quick and reliable connection to the ground line. A portable all-wave trap of adequate inductance, yet small dimensions, uses a powdered-iron core (Fig. 24). The core is so rated that an accidental connection to the power voltage will result in a cancellation of the inductance through over-saturation of the core and, consequently, in practically resistance-free grounding. The weight of the trap is approximately 20 kp. As has been previously explained, two traps are required for each construction site.

Fig. 24. Portable three-phase broad-band trap (Siemens A.G.)

The rating of portable wave traps is governed by the same considerations as apply to the traps permanently installed in the ground lead of the disconnect switches (see page 29). Objections based on national safety regulations have been raised in many countries to the use of portable wave traps as well as to the insertion of permanent wave traps in the ground lines. The usefulness of these types of traps becomes more than doubtful if such a trap was to be employed for grounding one three-phase system of a double line while the other system remains in operation [40].

3.3 By-Pass Circuits

By-pass circuits route the communication current around sub-stations or other points of separation in the power line. During the early stages of development, when antenna coupling was the only expedient, by-pass circuits were established by stringing an aerial wire in parallel with the two end sections of the power lines to be coupled. This type of by-pass proved to be just as inefficient as antenna coupling is by its very nature. Coupling capacitors, as they are used for line coupling purposes, have therefore been soon adopted also for by-pass circuits.

To safeguard operations, it is important that carrier by-pass circuits do not transfer any dangerous voltages from live power line sections to

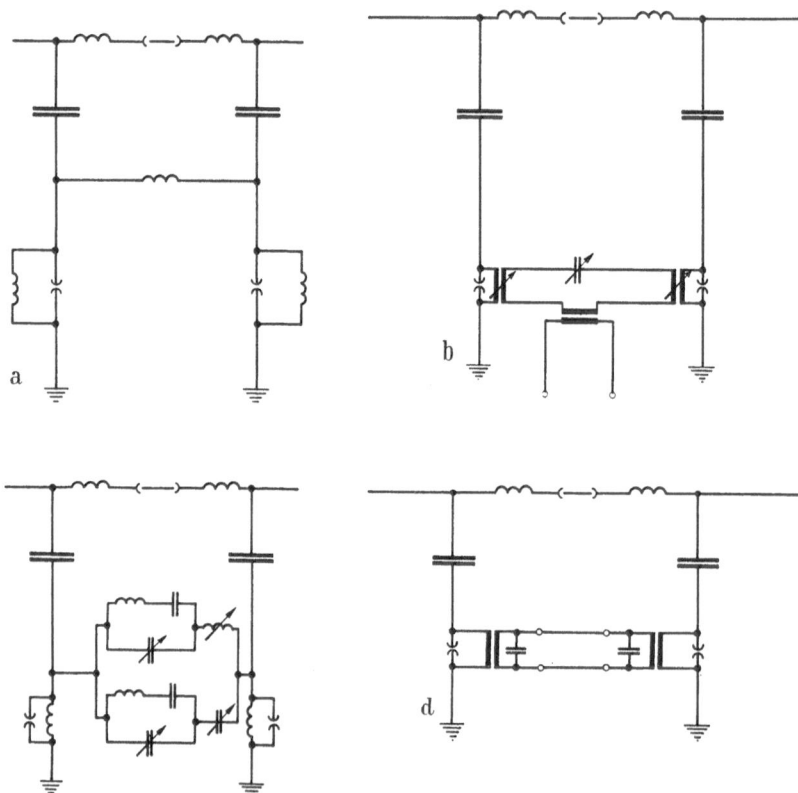

Fig. 25 a–d. By-pass circuits with phase-to-ground coupling.

a) By-pass with tuning coil between capacitors; b) By-pass with metallically separated arms and coupled telephone station; c) By-pass resonated to two carrier frequencies; d) By-pass with band-pass filters for a broad frequency band

a disconnected section. This is the reason why a separating point cannot be bridged simply by providing a capacitor.

The requirement for by-pass circuits excluding the possibility of an electric shock hazard are met if each capacitor is grounded through a sufficiently large inductance. In one of the earliest types of by-pass circuits (Fig. 25a) this precautionary measure had not yet been taken. This shortcoming was eliminated later when both arms of the by-pass were inductively coupled (Fig. 25b). Other by-pass circuits were resonated to the frequency pair to be passed (Fig. 25c). As in the case of resonated coupling media, quality of speech suffered from the fact that the frequencies are cut off at the limits of the voice band. In by-pass circuits of recent design (Fig. 25d) two normal coupling units, built as bandpass filters, are simply interconnected by a cable. The passband range of this type of by-pass corresponds to that of the usual line coupling arrangements.

Resonated by-pass circuits allowed only the two transmitted carrier frequencies to pass. An expansion of this circuit in order to pass more

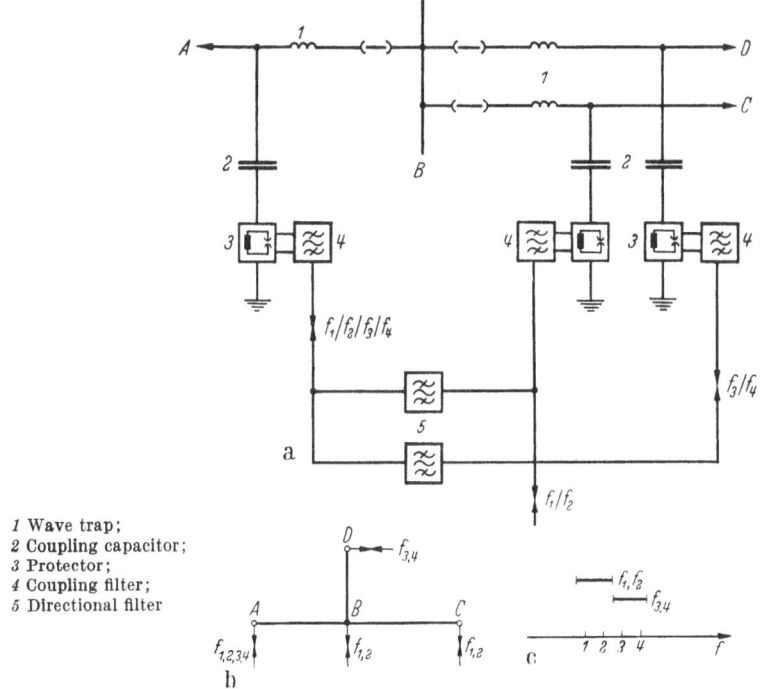

1 Wave trap;
2 Coupling capacitor;
3 Protector;
4 Coupling filter;
5 Directional filter

Fig. 26a–c. Three-way by-pass with additional bandpass filters for certain passband ranges
(Phase-to-ground coupling).
a) Basic circuitry; b) Transmission problem; c) Passband ranges of directional filters

than two carrier frequencies is a comparatively complicated and laborious job. An arrangement including bandpass filters is inherently suited to allowing a number of carrier frequencies to pass. This wide passband range is actually found desirable unless carrier frequencies have to be sent to or received from only one of the two directions in the station to be by-passed.

To avoid undue attenuation of carrier signals transmitted from the by-passed station in only one direction, or received there from only one direction, broad-band traps must be installed in either section of the power line.

Where resonant traps are exclusively used, additional traps are required also in the second line section. Preventing the transfer of frequencies not required in the other line section in the first place is a solution which is both more economical and more favorable as far as frequency assignment is concerned. This purpose can be accomplished with the aid of bandpass filters having different but overlapping pass-band ranges. However, to facilitate frequency assignment, it is now the normal practice to use "directional filters", which are inserted in the connecting line between the two coupling circuits and allow only the desired frequencies to pass (Fig. 26).

Another alternative is the "three-way by-pass". With the former tuning to resonance of by-pass circuits, this arrangement involved some tuning difficulties, which can be avoided if bandpass filter circuits are

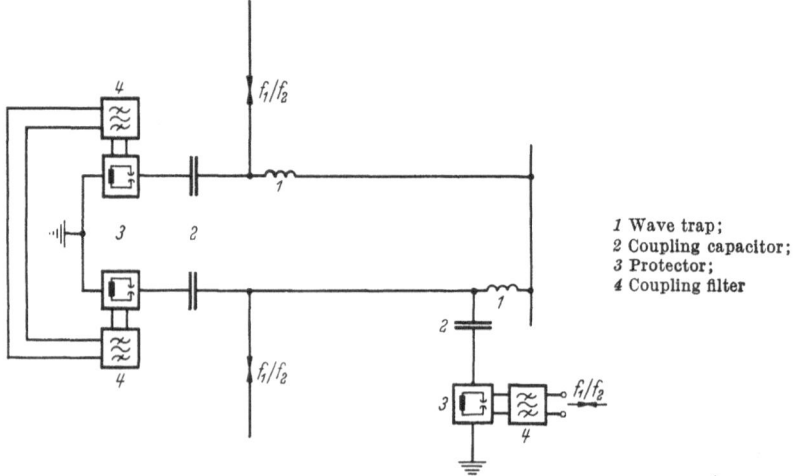

1 Wave trap;
2 Coupling capacitor;
3 Protector;
4 Coupling filter

Fig. 27. Three-way by-pass at a loop-in point

used. Three-way by-passes are rarely found. One-way by-passes predominate although a different passband range must in many cases be selected for the three directions to prevent unwanted dissemination of carrier frequencies in the power network.

A special variant of a three-way by-pass is used at points where the line is looped through the sub-station (Fig. 27) and where the carrier channel is to proceed beyond the loop-in point and, at the same time, is to be tapped for connection to the sub-station.

The separating point to be by-passed must not necessarily be a complete station. It might as well be an outdoor disconnect switch for interrupting the power line. Engineers occupied only with problems relating to power engineering are inclined to think that HF wave traps are here superfluous because there can be no question of HF current flowing to ground by way of the capacitances of bus bars or transformer windings. This opinion is only partly correct. The conditions for carrier frequency transmission are normal only as long as the disconnect switch

1 Disconnect switch; 2 Coupling capacitor; 3 Protector; 4 Coupling filter; 5 Auxiliary switch, 5 are closed as long as 1 is open, they are open when 1 is closed

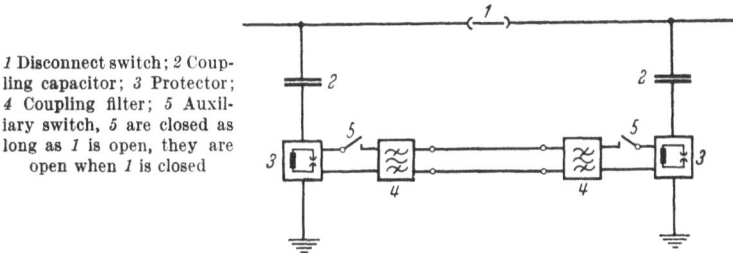

Fig. 28. Disconnect by-pass without wave traps.

is open in this disconnect by-pass, which differs from a conventional by-pass only in so far as no wave traps are provided (Fig. 28). If the disconnect switch is closed, the bandpass filters arranged between power line and ground are likely to introduce reflection losses which impair carrier communication. Auxiliary switches are therefore installed which cut out the coupling filters when the disconnect switch is closed, and thereby render the by-pass circuit ineffective.

3.4 Line Coupling in the Light of New Developments

Extensive investigations had been conducted during the first ten years of development in an effort to utilize the possibilities inherent in the power system for coupling and trapping purposes, and thus also for by-passes, in other words, to avoid the extra cost of line equipment for carrier communication. During the following thirty years, when the results of these investigations proved the impracticability of cost saving measures along these lines, the classical arrangement of a coupling circuit with power capacitors and a choke coil acting as wave trap has been universally adopted. Attempts have occasionally been made to classify the line equipment as belonging to the field of power engineering [11]. Nevertheless, communication engineers were untiring in their efforts to reduce the cost of carrier communication installations and to improve their performance characteristics. Although the results of these studies have so far not been turned to practical account, except for some isolated cases, it is certainly worth while to discuss some of the more recent efforts in this direction.

Plans to insulate the ground wires so that the line equipment would no longer have to be rated for the operating voltages and the operating current have already been mentioned (see page 15). Numerous systems of this type have been installed in a few countries [29, 44]. Economy is the prime motive for this type of systems. From a technical point of view, it is true that compromise solutions must then be accepted both as regards power engineering and carrier communication, but in most cases the disavantages of this compromise are not serious. It should be noted that the ground wire is occasionally insulated to reduce the power losses and not to establish a carrier frequency connection. In such a case the insulated ground wire will make a good communication path which requires no additional expenditure.

Another idea, based on pure technical considerations, has also been discussed earlier in this book. It provides for the installation of network decoupling circuits (see page 37 and Appendix 9.4). Such circuits could be used not only for establishing separate carrier communication networks in a metallically interconnected power network, i. e. for frequency conservation purposes, but also for decoupling a power line section at both ends so that a larger bundle of communication channels can be accommodated, say some 12 or 24 telephone circuits, as in the carrier systems of postal administrations or in radio relay connections. Without the use of decoupling circuits the larger portion of the carrier frequency range would be blocked for all of the remaining sections of the power network.

The same purpose, i. e. to provide multi-channel carrier communication on one line section and, at the same time, to make the entire frequency range available for the remaining sections of the mesh, is pursued in yet another way. This approach is based on the assumption that large channel groups are required chiefly along the right-of-way of power transmission lines, carrying 380 kV or even higher voltages. To reduce corona losses, these extra-high voltage lines are no longer made up of single conductors for each phase. Instead, two or four conductors are connected in parallel for each phase, so that the line may be said to consist of "bundled conductors". If means and ways are found to insulate the individual conductors of each phase between the stations at economically justified costs, and without impairing the power transmission characteristics of the line, and to reconnect them in parallel by means of wave traps only in the stations, it will be possible to provide multi-channel carrier paths without network decoupling circuits installed at the ends of the line. Fears that the carrier frequencies might be carried away into other sections of the network are hardly justified in view of the usual 40 cm separation between the conductors of a bundle. The mechanical and the electrical behavior of bundled conductors at power frequencies both during normal operation and in the event of trouble conditions has been investigated and as far as carrier communication is concerned, measurements have established the fact that these lines meet the expectations derived from theory.

Attempts have been made to apply carrier communication over power lines to fields outside the scope of electric system operation. Carrier telephony for farms is one of these applications [16]. Power lines erected for the distribution of electrical energy to a number of outlying farms (medium voltage lines of approx. 30 kV) are here used for public telephone service. This does not involve any new arrangements as regards line equipment. For train-to-wayside communication, which is another special application, a number of coupling methods have been employed, such as

a) antenna coupling to open-wire lines strung in parallel with the right-of-way (for non-electrified routes),

b) antenna coupling to a special carrier frequency conductor strung in parallel with the right-of-way (for electrified and non-electrified routes),

c) capacitive coupling to the collector of electric locomotives (for electrified routes).

An unusual application of line equipment has been mentioned in Australian engineering journals [43]. Referring to measurements carried out on a 500 kV experimental installation in the United States of America, suggestions are made to design the line equipment so that it will act both as coupling medium of carrier communication systems and also as a noise reduction circuit. This is to prevent radio interference voltages generated in a station from being radiated from the power lines. Interference above the carrier frequency range (up to 10 MHz) has thereby been reduced by 40%.

4. Properties of Power Lines in the Carrier Frequency Range

Carrier communication over power lines started with equipment borrowed from wireless telephony. The previously mentioned coupling antenna is one of the most characteristic examples. Transmission was believed to be a space radiation process with wave propagation concentrated along the power line route. "Line-bound HL transmission" or similar terms were used at that time to distinguish this mode of operation from wireless radio frequency transmission. There is no essential difference, however, between the transmission of electrical energy with commercial AC currents of $16^2/_3$ radio frequency or 50 Hz and the transmission of HF currents at frequencies ranging between 15 kHz and 500 kHz. Current and voltage of the line produce an electromagnetic field between the conductors and the energy is transmitted through this field [6]. It is true, however, that line properties vary greatly with the frequency. Energy losses, in particular, increase rapidly at higher frequencies. The attenuation of the line is a measure of energy losses (Appendix 9.5). It is a function of loss resistance and the characteristic impedance of a line.

Transmission line theory may be used as a basis for calculating these two variables and, consequently, the line attenuation. What must be known are the geometric dimensions, such as separation of conductors, clearance between conductors and ground, diameter of conductors, and the material of which conductors are made. The fact that the non-coupled conductors are also engaged in carrier frequency communication, due to capacitive and inductive interaction with the coupled conductors, is not sufficiently appreciated in conventional calculation methods. The influence of the ground on the transmission process must also be considered.

During the past years the theoretical investigations into the transmission process have found increasing interest, in particular with respect to power lines carrying extremely high operating voltages, such as 500 kV or 750 kV. Previous experience as regards the suitability of these lines for carrier frequency communication was not available. The "Modal Analysis" [33, 42] supplied the base for a rather accurate calculation of the transmission conditions on lines with few transpositions and with turrets whose dimensions considerably exceed those previously known.

All these calculations are a rather toilsome approach. Even a generally applicable table for the different types of lines, if such a table would exist, could not be applied without some difficulty. This is due to the fact that too many variants in line construction would have to be considered. Measurements have therefore been carried out over many years on a wide variety of lines and approximate values have been determined to provide a guide for covering the characteristics of power lines together with those of the coupling circuits. Exhaustive theoretical examinations are only made in the case of extra-high voltage networks.

Data dispersed over a large number of publications are frequently found to differ considerably. A critical examination, survey, supplementation and final editing of all data results in "characteristics", the numerical values of which should be known for designing and installing carrier communication systems[1]. The most important characteristics can be represented by generally applicable guiding values. These can be used in most cases for designing communication systems and forecasting their performance without the necessity of previous measurements. In some systems additional characteristics must be taken into account. Their

---

[1] The following passages in sections 4 and 8.1 through 8.3 have been taken, for the most part, from a study prepared by E. ALSLEBEN [13].

values vary considerably so that only rather wide limits can be stated. Since they differ too much with each specific case and the operating condition of the power system, they must be determined by measurements. Unless such measurements can be carried out during the planning stage, they should be made when the system is being put in operation, to ensure satisfactory service under all possible operating conditions.

The signal-to-noise ratio obtained at the receiving station is a deciding factor for the quality of a communication circuit (Appendix 9.5). The attenuation of the useful signal and the magnitude of the interference level are important characteristics of the power system, and therefore of great interest to the communication system designer.

The characteristics which essentially contribute to the attenuation and are of importance when rating a transmission system for minimum attenuation are determined by the set-up of the power network and the design of the coupling circuits. The characteristics which can be given for the interference level also depend on the network configuration apart from the actual causes of interference, such as power system operation. atmospherics, and external communication circuits. Weather conditions. too, have a marked effect on a number of characteristics.

To obtain the maximum output from a signal generator, the generator is matched to the load in communication engineering, contrary to the practice adopted in power engineering. The impedance of the transmitter is matched to the line impedance, and the latter to the impedance of the signal receiver at the end of the line.

If messages are transmitted over conventional communication lines, impedance matching presents no major problem so that matching losses will not constitute any appreciable part in the total attenuation. Attenuation is here due chiefly to energy losses along the line. The decisive factor for impedance matching is the characteristic impedance of the line. It may be expressed as the input impedance of the infinitely long, or of the characteristically terminated and, hence, reflection-free, short. homogeneous line.

In signal transmission over power lines, losses due to mismatching of the individual components of the transmission path (mismatch losses), and to leakage of carrier currents into the connected power facilities (shunt losses, see page 55) must be added to the line loss proper. The input impedance of a short line, which is not terminated reflection-free, exhibits maxima and minima in dependence on the frequency. The longer the line, i. e. the more the influence on the terminating impedance decreases as a result of increasing attenuation, the more will the input impedance approach the characteristic impedance as a limit value.

The characteristic impedance of a transmission system made up of several conductors and ground may be pictured as being composed of part impedances effective between the individual conductors, and between the conductors and ground. The carrier energy applied to the system by way of the coupling circuits is distributed between the conductors on the one hand, and conductors and ground on the other hand, in dependence on the interconnection of these partial characteristic impedances and any terminating impedances which may be inserted. Different conditions result for the two most usual coupling methods, phase-to-phase coupling with balance to ground on the one, and phase-to-ground coupling on the other hand.

With phase-to-phase coupling, the major portion of the energy transmitted will be found between the coupled conductors. It is propagated between these conductors in the form of an electromagnetic wave. The portion between the conductors and ground is comparatively small. If complete balance is achieved, the third conductor will not carry any carrier current. Even in the usual, less accurately balanced arrangements, the third conductor, like the ground wire, takes little part in message communication. Assuming ideal trapping media in the two conductors used for line coupling, leakage of HF energy into the substation could be entirely prevented. At the end of the line, conditions are reversed. The total of the energy between the two conductors can be delivered to the receiver reflection-free. With phase-to-phase coupling, the characteristic impedance between the two coupled conductors, which is independent of the condition of the station, can therefore be measured as the input impedance of the line, to which transmitter and receiver must be matched to ensure satisfactory transmission conditions.

A different distribution of energy is obtained in the phase-to-ground coupling case. Part of the carrier energy is lost as it flows into the station by way of the two untrapped conductors. A substantial part of the remaining energy, propagated as an electromagnetic wave, is distributed between conductors and ground, while the rest is distributed symmetrically between the coupled conductor and the two non-coupled conductors. The latter are practically at the same carrier potential (at exactly the same potential in case of complete balance) and each of them carries half the current of the coupled conductor. As a result of the high ground losses, the portion between conductors and ground suffers much greater attenuation than the portion between the conductors which, at a certain distance from the feed-in point, is practically all that is left of the original energy, and which is then propagated under conditions similar to those obtaining with phase-to-phase coupling. The conditions at the end of the line correspond to those prevailing at the beginning of the line. The receiver is connected between the coupled conductor and ground rather than between the coupled conductor and the two non-coupled conductors, even through this arrangement would appear to be in conformance with the incoming wave. With this type of connection, the line can therefore not be terminated reflection-free by matching the receiver. Only part of the incoming energy will actually reach the receiver. The rest ist partly lost in the station and partly reflected. In the case of phase-to-ground coupling, the input impedance will therefore not depend on the line alone but also on the input impedance which the station presents to the two non-coupled conductors. For a sufficiently long line, with no reaction from the output of the line to the input, the input impedance would assume a limit value which depends on the station and its switching condition. There is some justification in referring to this limit value as to the characteristic impedance with phase-to-ground coupling, and it appears to be appropriate to match the transmitter and the receiver to this value in order to achieve minimum attenuation. Nevertheless, reflection-free termination cannot be secured with this type of matching.

Whereas in the case of phase-to-phase coupling the attenuation of the power line is chiefly due to propagation losses and therefore increases in proportion to the length of the line, a "phase-to-ground additional attenuation" must be added in the case of phase-to-ground coupling. This incremental attenuation is independent of the length of the line and takes account of the wave, generated at both ends between conductors and ground, whose energy is practically lost if the line exceeds a certain length. Another factor covered by this additional attenuation is the energy flowing into the station. In some publications the data given on attenuation with phase-to-ground coupling have not been resolved into these two constituent components. Therefore, they cannot be readily applied to lines of similar design but of different length.

In conventional communication circuits, matching to the characteristic impedance can be performed so as to avoid a noticeable mismatch loss and an incremental attenuation. In the case of power lines, however, an incremental attenuation introduced by the coupling circuit and the leads to the communication equipment must be taken into consideration.

The impedance of a station has an appreciable influence on the incremental attenuation occuring with phase-to-ground coupling. But even in the case of phase-to-phase coupling its influence must not be entirely neglected because, for reasons connected with power system operation, the wave traps intended to eliminate this influence cannot be given the rating necessary to obtain a device which would perfectly suit the needs of carrier communication. This fact is borne out by the impedance presented by a station, which has the effect of a shunt loss.

Another characteristic, the transit loss of a station, is of interest in all cases where a circuit is to continue over a second line on the other side of the station, or where the same carrier frequency is to be used for different circuits in the same power network.

Power cables are occasionally used for interconnecting overhead line sections and as entrance cables to stations. They may also be employed for carrier communication. The characteristic impedance and attenuation of power cables must then be known.

Two or more mismatch points in a line are annoying in particular because, considering the usual distances and line losses, reflection will cause standing waves which result in cyclic and frequency dependent reflection losses. These, in turn, are likely to result in an objectionable frequency response within the bands used for the individual communication circuits. They are likely to occur not only at the ends of the line but also at mismatch points which are due to peculiarities in line configuration, at branch-off points, tie lines, or other inhomogeneities of the line. Special attention should therefore be given to these factors when engineering a communication circuit for operation over power lines [12].

The following is a compilation of numerical values for the various characteristics. These values have been taken from a number of publications dealing with measurements and calculations.

4.1 Characteristic Impedance of Overhead Power Lines

For the frequency range of interest, the characteristic impedance of a two-conductor power line having a sufficient ground clearance is given by the expression

$$Z = 120 \ln \frac{d}{r} \, \Omega \left(= 276 \log \frac{d}{r} \, \Omega \right)$$

where

d = separation of conductors, and
r = radius of conductors.

For phase-to-phase coupling, this formula will yield results closely approximating the characteristic impedance. The latter represents a real value which is independent of frequency. The actual figures are slightly lower due to the influence of the ground and the influence of adjacent conductors. However, these deviations are normally negligible.

The line configuration is of minor importance because only the logarithm of the large quotient d/r enters into the formula. Besides, lines rated for a higher operating voltage have normally a larger separation and conductors of a larger diameter so that the ratio of d/r increases approximately with the root of the voltage.

Substantially lower values are obtained only with the bundle conductors of lines in the high voltage class. For calculating the characteristic impedance, the radius r must here be replaced by an effective radius r', which is by far larger than the radius of the individual conductor:

$$r' = r \left(k\, \frac{s}{r} \right)^{\frac{n-1}{n}}$$

where

s = separation between adjacent conductors within the bundle

n = number of conductors of a bundle

k = a constant derived from n:

n	2	3	4	5
k	1	1	1.12	1.27

Where more than two conductors are used to form a bundle, the effective radius of the bundle will increase and the characteristic impedance diminish.

With phase-to-ground coupling in a three-phase system, a very elaborate theory permits the input impedance to be calculated for any length of line, taking into account the influence of the impedance in which all phases are terminated [33, 42]. Such accurate calculations are not necessary, however, for system planning.

Initial plans may be based on the following approximate values:

Phase-to-phase coupling:	Range	Planning figure
Single conductor	650–800 Ω	700 Ω
2-conductor bundle	500–600 Ω	500 Ω
4-conductor bundle	420–500 Ω	500 Ω
Phase-to-ground coupling:		
Single conductor	350–500 Ω	400 Ω
2-conductor bundle	250–400 Ω	325 Ω
4-conductor bundle	220–350 Ω	325 Ω

4.2 Attenuation of Overhead Power Lines

For the frequency range employed in power line carrier communication the loss of a two-conductor line, having a sufficient ground clearance, can be calculated from the loss resistance and the characteristic impedance of the line

$$a = \frac{R}{2Z}.$$

The skin effect causes R and, consequently, a to rise with the root of the frequency.

The attenuation of a power line with phase-to-phase coupling is higher than the attenuation of a two-conductor line having the same dimensions, because ground losses must be added to the actual line loss and the former are frequently higher than the line loss proper. This is due to the small clearance between conductors and ground. Depending on the line configuration and the location of the coupled phases, this clearance is only one to four times the separation between the conductors. This results in high eddy current losses and, at higher frequencies, in increased dielectric losses in the ground. These losses lead to an increase in the loss resistance of the power line. Furthermore, they cause the resistance and the attenuation of the line to increase almost in direct proportion to the frequency in contrast to the loss values found for a line which is not influenced by the ground. This holds for lines carrying voltages over about 60 kV. In lower-voltage power lines, the conductors have a smaller cross section and the losses in the conductors are here the dominating contribution to the total loss.

In the lower-voltage case, a high humidity content of the air will cause leakage losses across the insulators and this results in a higher attenuation, whereas high-voltage lines with their larger insulators are hardly affected. A slight increase in attenuation might also be noticed on rainy days if the conductors are excessively contaminated.

Hoarfrost and ice build-up tend to multiply the attenuation figures. This fact should be duly considered in areas where such conditions are frequently encountered. Excessive ice formation on a line in Norway raised the attenuation to a figure fifteen times higher than the normal values at a frequency of 105 kHz [30]. Under normal conditions, one is safe in assuming that an excessive increase in attenuation is restricted to particularly exposed spots and will not occur over the entire length of the line. Similar values [37] have been measured in Switzerland, again under extreme weather conditions. Sufficiently conclusive measuring results are as yet not available on the frequency versus loss characteristic under conditions of hoarfrost and ice formation. In theoretical examinations the increased attenuation has been attributed to dielectric losses in ice and, up to a range of 200 kHz, the attenuation was found to rise approximately in proportion to the frequency.

If all experience data are gathered together to determine the line attenuation under fair weather conditions, planning values as shown in Fig. 29 will be obtained for lines between approximately 60 kV and 400 kV. It is evident that the line attenuation depends greatly on the ratio d/h. The separation d between the conductors is to be understood as the mean value, which takes into account any transposition of lines. In the phase-to-ground coupling case, it refers to the mean spacing between the coupled conductor and the two adjacent conductors. The height h represents the mean ground clearance at the towers.

With phase-to-ground coupling, this attenuation value is increased by the additional attenuation of phase-to-ground coupling. Its magnitude is a function of the input impedance which the station presents to the non-coupled conductors. It has been determined by calculation and field tests for the two extreme cases of short-circuit (grounding) and open circuit (disconnection) and was found to be about 2.2 db and 5.7 db, respectively. For transmission over several line sections, these values must be added for each section.

Normally, no line traps are provided for the non-coupled conductors. The incremental attenuation then depends, within the stated limits, on

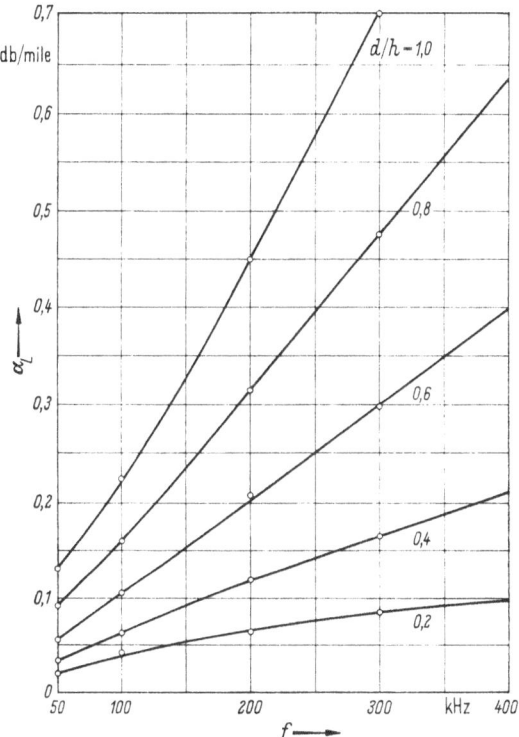

Fig. 29. Planning values for the line attenuation
at operating voltages between 60 kV and 400 kV

the conductor-to-ground impedance of the two conductors entering the station and, consequently, on their switching condition. A planning value of 3.5 db is usually taken for one line section.

Where a capacitive voltage transformer is available for each of these conductors, the two non-coupled conductors can readily be shorted with respect to the HF current. A minimum of additional attenuation is attained in this way and the dependence on the switching condition is diminished. A planning value of 2.2 db can be assumed in this case for one line section. The open-circuit case of the two non-coupled conductors (disconnection of line without grounding by way of switches or capacitive voltage transformers) need not be considered here because, although the attenuation increases, the noise level is reduced, and any increase in attenuation is normally compensated by the automatic volume control of the carrier terminals.

4.3 Coupling Loss

It is common practice today to use broadband coupling filters. These are intended to match the impedance of the communication equipment as exactly as possible to the line impedance. With the customary circuit arrangements, accurate impedance matching can be accomplished for two frequencies only. A mismatch loss will occur for any other frequency. Its maximum value, at a given capacitance of the coupling capacitor, depends essentially on the frequency range to be used for transmission. Coupling circuits are normally so rated that the mismatch loss introduced by the filter will not exceed 1.3 db. To this value should be added a small loss in the coupling capacitor and in the coils of the filter circuit, which amounts to a total of approximately 0.4 db in the higher frequency range. Thus, the coupling circuits proper will present an attenuation of up to 1.7 db. Another factor, also to be considered in this connection, is the attenuation of the feeder line, a cable in most cases, which amounts to approximately 3.2 db/mile at a frequency of 300 kHz, using the usual paper-insulated symmetrical cables which are manufactured specifically for establishing connections to the coupling circuits.

4.4 Impedance Presented by a Power Network Station

The impedance presented by a power network station is of particular interest if means and ways are to be found to avoid wave traps. As a rule, this impedance is made up of several components and it is more or less a function of the frequency. It is composed of the capacitive impedance of bus bars and switches, the impedance of measuring transformers (also capacitive in most cases), and the impedance of the power transformers, which may again be capacitive, but which will frequently vary with the frequency between capacitive and inductive. A distinction can be made between:

a) Switching stations in the strictest sense of the word (including no power transformers) which, depending on their size, constitute a more or less significant capacitance.

b) Small transformer sub-stations, equipped with only one power transformer. Their switching condition varies infrequently, and their impedance is given by the bus bars and by the transformer. In some cases the impedance may be high and sufficiently independent of the frequency so that wave traps need not be provided.

c) Large sub-stations, including a number of transformers, whose switching condition changes rather frequently. The impedance is often closely tied up with the frequency and its value will range from high to low. Since these values vary greatly with switching, the installation of wave traps is here imperative.

The bus bar capacitance can be roughly calculated if the dimensions are known. For an existing station it will be more convenient to have it determined by measurement. The values found will range between 0.5 nF and 10 nF, depending on the size of the station. The transformer impedance, if it is capacitive, will range between 1 nF and 5 nF, depending on the size of the transformer. There is not much sense in trying to determine the impedance of a large station with the aid of individual values, obtained either from statistics or by measurement, or to measure it as a total value. The results of such computations are unreliable and the measured value depends too much on the prevailing switching condition.

4.5 Shunting Loss of a Power Network Station

Wave traps are intended to prevent the HF energy from traveling away into the connected power network stations. Their blocking efficiency is limited, and a certain amount of shunt loss must be expected which varies with the switching condition of the station. Nevertheless, guiding values can be given also for this type of transmission loss (Applendix 9.2).

The usual practice is to assume an average value of 1.3 db for the incremental loss introduced by a power network station for which wave traps are provided.

4.6 Transit Loss of a Power Network Station

Satisfactory communication is in most cases no longer possible if signals are transmitted directly through intermediate power network stations. Small stations may be an exception to this rule. A high transit attenuation, i. e. a minimum of crosstalk from one line to the other lines radiating from the station, is desirable for two reasons.

Firstly, the carrier energy passing through the station to other lines prevents the same frequency from being used again for other communication circuits, unless these circuits are separated from the point of entry of the carrier energy by a large geographical distance. This fact tends to complicate the task of frequency assigners.

Secondly, in the case of a by-pass circuit, the carrier energy transferred to the other line by way of the station is added to the carrier frequency taking the legitimate path, i. e. the by-pass circuit, with an undefined, frequency-dependent phase position. The result may be a certain increase in or weakening of the carrier energy, the deciding factor being the difference in attenuation between the two paths. If the by-pass circuit includes a repeater for the incoming signals, too low a transit attenuation might involve the necessity of frequency conversion to avoid annoying feedback.

It is not possible to establish generally applicable guiding values for the transit loss because they are too closely associated with the type and the size of a station. Furthermore, their magnitude depends on whether wave traps are provided for all or only for the coupled conductors. Another important factor is the line arrangement, i. e. whether the outgoing lines are run side-by-side on the same tower structures, or whether they radiate into different directions.

Calculations may be based on values ranging between 9 db and 40 db. To be on the safe side, it will be advisable to determine the actual values through measurements.

4.7 Characteristic Impedance and Attenuation of Power Cables

As in the case of overhead power lines, the characteristic impedance and the attenuation of the power cable to be used for carrier communication must be known. Both these values deviate considerably from those found on overhead lines. Since they differ with the type of cable used, they will have to be determined by measurement in each specific case.

Cables carrying higher voltages are normally constructed so that the individual phase conductors are surrounded by a conducting sheath which is connected to ground. This permits direct transmission between one conductor and ground. The characteristic impedance lies somewhere between $25\,\Omega$ and $35\,\Omega$ for phase-to-ground coupling, and these values may be doubled for the phase-to-phase coupling case. Quite generally, it may be said that the attenuation of cables is about ten times higher than the attenuation on overhead lines.

Cables are sometimes used for the lead-in section between power line and power network station. The transition from the overhead line to the power cable results in a substantial mismatch loss across the cable input. It will therefore be advisable to arrange the line equipment at the transition point as, otherwise, a special matching circuit will be required between power line and cable. This applies also to the case where cable sections are inserted along the right-of-way of the power line. The transit loss of stations using an entry cable by far exceeds the loss experienced in stations directly connected through the overhead lines [26].

4.8 Attenuation of a Complete Transmission Section

Power lines cannot normally be assumed to constitute a completely homogeneous circuit for communication purposes. A minor mismatch, as may be caused by the transposition of the power conductors, will hardly cause any trouble. A serious mismatch will, however, occur at the previously mentioned points of transition between an overhead line and cables, at branch-off-points, and at spur line connection points. The result is an excessive incremental attenuation and frequency response caused by reflection. Special measures must here be taken to ensure satisfactory carrier transmission.

There are no set guiding rules on these cases so that calculations or measurements have to be carried out to determine in each specific case whether signal transmission is impaired and whether special measures become a necessity.

As can be seen from the foregoing, the attenuation a of a complete transmission section is made up of various components

$$a = \alpha_l \, l_l + (\alpha_{add} + 2\alpha_{cp} + 2\alpha_{st}) \, \alpha_{cbl} \, l_{cbl}$$

where, with the previously mentioned values inserted,

α_l	= line attenuation per unit length of line (Fig. 29)	
l_l	= length of line	
α_{add}	= additional attenuation with phase-to-ground coupling	= 3.5 db or 2 db
	additional attenuation with phase-to-phase coupling	= 0 db
α_{cp}	= attenuation of coupling circuit	= 1.7 db
α_{st}	= attenuation introduced by the power network station	= 1.3 db
α_{cbl}	= attenuation of HF entry cable (depends on make and frequency), for an example frequency of 300 kHz	= 3.2 db/mile
l_{cbl}	= length of HF entry cable	

If the carrier frequency is transmitted over a number of tandem-connected power line sections, the same formula can be applied for determining the attenuation between the two terminal points. To the result must then be added the attenuation of the by-pass circuits in each intermediate power network station (see also level diagram, Appendix 9.5).

As long as a sufficient separation is provided between the carrier frequencies used on two transmission paths which radiate into different directions from a power network station, it will be immaterial whether the two carrier terminals are connected to the same or to different conductors. Unwanted HF currents of a different frequency are at any rate absorbed by the separating filters installed in the carrier terminals. Nevertheless, the two carrier terminals should be connected to different conductors, if possible, to allow for a closer spacing of the frequencies, where this is required. This tends to lessen crosstalk because, in view of the metallic separation, coupling between the two conductors is restricted to the magnetic field surrounding the conductors.

Thus, different conductors are coupled in an effort to maximize the crosstalk attenuation between carrier districts converging in a power network station. Where a carrier circuit is through-connected by means of a by-pass circuit, a maximum transit attenuation is again desired to avoid the danger of mutual cancellation. If the by-passed station is equipped with a non-conversion repeater, a high transit loss is desired to prevent feedback.

4.9 Noise Level

Power transmission introduces high noise voltages in the carrier circuits. Their interference power by far exceeds that found in postal communication circuits [25]. (According to a CCITT definition, the term "interference voltage" [interference level] refers to an external AC voltage interfering with the communication channel.)

In the past, power lines and power equipment were in most cases built without giving thought to the disturbance which they might cause in carrier frequency communication and in radio broadcasting. In the modern extra-high voltage transmission systems rated for 380 kV (and higher), special noise-reducing measures have been taken with regard to power equipment, lines and fittings. This was done not only with a view to improving radio reception and carrier communication over the company-owned transmission system, but also to minimize energy losses within the power transmission system.

The behavior of newly developed equipment for extra-high voltage systems is carefully studied in the large test fields of the manufacturers and in the experimental plants set up in the various countries before this equipment is released for general use.

The noise voltages generated and occurring in power transmission systems can be classified, as to their nature, into two categories. Noise voltages of the one category are present at all times. They are generated by discharges across insulators and line fittings, and by discharges on the lines proper. The noise voltages of the second category are of a temporary, impulse-type nature. They are caused by switching processes and atmospheric discharges.

Noise voltages of the first category normally produce, at low operating voltages, a non-uniform frying noise. At higher voltages a more uniform, rustling noise appears which is due to corona discharges. It is composed of a multitude of independent individual discharges along the line, particularly during the crest values of the positive half cycles. The corona noise thus varies periodically with the power frequency and its multiples, in particular with the third harmonic. When the voltage level rises during off-peak periods or when the dielectric conditions around the conductor change (rain, fog, hoarfrost), the noise voltage increases. The noise spectrum produced by corona discharges within the carrier frequency range has equal amplitudes up to about 1000 kHz. Beyond this limit, the amplitudes subside slowly. Another phenomenon, appearing in addition to the independent corona noise voltage, is "corona modulation".

The periodic variations of corona noise cause the transmitted carrier signals to be modulated in the rhythm of the power frequency. Each useful frequency is accompanied by a noise spectrum. Corona modulation is likely to affect double-sideband transmissions where the carrier has a greater amplitude than the sidebands so that its noise spectrum will also have a greater amplitude. This is all the more annoying in speech transmission, because the carrier and, consequently, its noise spectrum are present also during silent periods.

The following approximate noise levels must be expected in a 2.5 kHz frequency band under unfavorable weather conditions:

- 35 db for 110 kV lines
- 17 db for 220 kV lines
- 8 db for 380 kV lines

These guiding values apply to the power line proper. As far as the noise level at the input of the carrier terminals is concerned, a higher attenuation of the coupling circuit and of the entry cable may be taken into consideration.

A noise level lower than the guiding values may be expected during dry weather. It may be higher with wet weather and excessive pollution of lines in industrial areas.

The noise level on extra high voltage lines increases only slowly compared with 220 kV lines, because lines carrying more then 220 kV are made up of bundled conductors to reduce the field strength around the wires and, consequently, the corona losses. This measure, originally adopted to improve power transmission, benefits also carrier communication.

Noise voltages of the second category, i. e. impulse noise generated by flashovers, atmospheric discharges, and by the operation of power switches results in a very broad noise spectrum which affects the entire working range [17, 19]. During the recent past, this type of noise has been given closer attention [25, 14] because modern high-speed protective relaying terminals are required to operate reliably also in the presence of these impulse noise voltages (see page 127).

Accurate data on the magnitude of the interference level contributed by broadcasting transmitters and by other radio transmitters cannot be given [20]. If systems are planned with a frequency in a region where interference is most likely to occur, the effect of such interference should be determined by measurements. If interference is caused by radio transmitters, a lower noise level may be expected for phase-to-phase coupling than for phase-to-ground coupling [22]. This is also true in those cases where the interferers are carrier circuits operated at the same frequency over other line sections.

5. Characteristics of Power Line Carrier Channels

The limits of the repeatedly mentioned frequency range from 15 kHz to 500 kHz, which is made available for carrier communication over power lines, are determined by physical and economical conditions, partly also by administrative regulations.

The lower limit of this frequency range is fixed by technical considerations connected with the rating of coupling circuits and wave traps. Taking the usual coupling capacitances of 2200 pF to 4400 pF as a basis, coupling circuits having the passband range desired for multichannel coupling can still be designed, but the passband range of the coupling filter is much smaller at low frequencies than at high frequencies. The blocking efficiency of all-wave traps decreases with the frequency, and the remedy of increasing the inductance of the trap is limited for economical reasons. The VF operated centralized remote control systems, it is true, operate with still lower frequencies on power lines [24]. However, a much higher send power (up to several kW) is used which, contrary to carrier communication, is fed in in series with the power voltage. No traps are used.

The attenuation of a transmission line sets the limit to the upper frequency range. It increases rapidly with the frequency. This applies in particular to the incremental attenuation caused by hoarfrost. Low carrier frequencies are therefore preferred for carrier communication over power lines. When calculating the transmission range at a frequency of 500 kHz, using the send level of conventional carrier terminals and the usually encountered noise level as a basis, it will be found that only a comparatively short distance can be spanned, even if phase-to-phase coupling is adopted. The incremental attenuation caused by hoarfrost is high at carrier frequencies above 300 kHz compared with the attenuation measured during normal weather conditions. With frequencies ranging between 300 kHz and 500 kHz, it is therefore impossible to establish connections which ensure reliable operation also under hoarfrost conditions. This fact must always be borne in mind in the planning stage.

In some cases still higher frequencies are used for operation over power lines (approx. 1000 kHz). These frequencies are, however, no longer intended for message communication[1] but for fault location. HF pulses of high energy and lasting only a few microseconds are here transmitted. They are reflected from the faulted point and the delay time of the signal provides an indication as to the distance at which the trouble spot is located.

The lower portion of the 15 kHz to 500 kHz range lies below the range of long-wave radio broadcasting, but still within the range of carrier communication over the open-wire lines of postal administrations. Accommodated within this power line carrier range are some frequency positions assigned to the radio transmitters of air traffic and maritime communication services, and to transmitters of international news services. The rapid development in interconnected power system operation compelled the power companies to investigate the possibility of mutual interference between the frequencies employed in their communication networks, on the one and those used within the same range by a large number of radio services on the other hand [20, 22]. It was found that, while certain precautions must sometimes be taken during the planning stage to prevent carrier communication systems from being affected by interference, there is no danger of external radio receivers being affected by the operation of the carrier terminals of the power companies. The fact that some radio receivers installed in the immediate vicinity of power lines are affected by interference is not of decisive importance, considering the public interest served by the power plants. The possibility of poor radio reception, small as it is, can be further decreased by using radio receivers having a sharper selectivity, and by a suitable design of the carrier communication system.

In most countries the assignment of frequencies to the various communication services is the responsibility of the postal administrations so that any disorder in this respect is precluded. The frequency range which postal administrations make avialable for power line carrier

---

[1] Plans, based on suggestions by GOUBAU, to accommodate dense groups of communication channels at frequencies around 300 MHz, are here not dealt with because they will hardly be realized in the foreseeable future.

communication differs for the various countries, as can be seen from the table below:

Australia	152 kHz to 450 kHz	Japan	10 kHz to 450 kHz
Czechoslovakia	40 kHz to 500 kHz	Norway	44 kHz to 156 kHz
Germany	30 kHz to 490 kHz	Sweden	40 kHz to 500 kHz
France	40 kHz to 300 kHz	Switzerland	40 kHz to 300 kHz
Italy	50 kHz to 392 kHz	USA	30 kHz to 200 kHz

If the problem of frequency assignment is attacked in a cooperative spirit both by the power companies and the postal administrations, there will not be any need for barring certain sectors within the assigned frequency range for technical reasons.

5.1 Transmission Systems

During the earlier stages of the development, amplitude modulation was the only choice. Speech transmission made up the bulk of the traffic. The voice band transmitted took up the range between 300 Hz and 2400 Hz. Modulation of the carrier with this frequency band results in two sidebands in the carrier range. A bandwidth of 5 kHz is sufficient for transmitting these sidebands (Appendix 9.6). The total range available has been divided into contiguous frequency positions of 5 kHz width, with the nominal frequency located in the center. The rapidly growing requirements for new carrier communication circuits led to a shortage of frequency positions. Attempts were made to overcome this impasse by introducing single sideband (SSB) transmission (Appendix 9.6). The frequency band required for speech transmission is thereby reduced to one half, i. e. to 2.5 kHz. In other words, it became possible to accommodate twice as many speech circuits as before within the available frequency range. In Germany, where many double side band (DSB) carrier terminals were operated in a 5 kHz allocation scheme, the SSB systems were designed so as to fit into this 5 kHz scheme. Other countries proceeded along different lines and adopted DSB equipment for an 8 kHz, and SSB equipment for a 4 kHz allocation scheme.

Frequency modulation (Appendix 9.6) is not the appropriate way to conserve spectrum space, i. e. to obtain any advantage over amplitude modulation and DSB operation in this respect. In view of the frequency shortage and the resultant frequency assignment problems, a frequency modulated carrier transmitted over power lines should not take up more space than an amplitude modulated carrier with double sideband transmission. With a transmitted VF band of up to 2.4 kHz, the permissible frequency swing will be $\pm$ 2 kHz. This frequency swing affords a certain signal-to-noise advantage over amplitude modulation with double sideband transmission. Although a larger frequency swing would result in a still better signal-to-noise ratio, it cannot be readily justified as it would involve the necessity of providing a wider frequency band and thereby thwart all frequency conservation efforts.

Summing up, three different transmission methods are now employed for carrier communication over power lines:

a) Amplitude modulation with double sideband transmission,
b) Amplitude modulation with single sideband transmission,
c) Frequency modulation.

Investigations have been conducted to find out whether the SSB transmission method can be employed to advantage in connection with frequency modulation [28]. Contrary to amplitude modulation and single sideband transmission, it is here not possible to suppress the carrier at the transmitting station and to add it for demodulation at the receiving end, because its frequency affects the shape of the wave. As a result, the carrier cannot be omitted in frequency-modulated transmissions. Suppression of the carrier or of a sideband would result in excessive distortion and intolerable intermodulation products.

Frequency modulation permits the design of carrier terminals requiring the same frequency space as DSB equipment. However, the send power necessary for spanning the same distance is lower and this means a saving in cost (see page 121). Using the same send power, a greater transmission range could be obtained, although this range is not larger than that of SSB terminals, which occupy only half the frequency space. A rough comparison of the three types of equipment can be given by stating relative figures of merit (Fig. 30).

Type of equipment	SSB	DSB	FM
Cost figure	3	2	1
Send power figure	4	4	1
Range figure	3	1	1
Frequency figure	1	2	2

Fig. 30. Comparison of the three conventional types of transmission terminals with relation to cost, send power, transmission range, and frequency requirements

5.2 Bandwidth and Frequency Assignment

According to CCITT standards, a frequency is effectively transmitted if the overall equivalent at the relative frequency does not exceed the overall equivalent at a frequency of 800 Hz [34] by more than 8.7 db. Furthermore, a minimum value of the overall equivalent has been established which must be observed irrespective of the frequency. Considering the attenuation conditions corresponding to the CCITT recommendation, originally intended to cover telephone lines (see Fig. 31), a distinction is made between two-wire lines for short-haul communication and the higher-grade four-wire lines for long-haul communication. The bandwidth of a telephone circuit is therefore defined by the difference between the two edge frequencies, which satisfy the previously mentioned 8.7 db requirement. Accordingly, it is 2.1 kHz or 2.3 kHz.

During the years around 1938 the Study Committees of the CCITT suggested the use of wider voice bands, to improve the quality of trans-

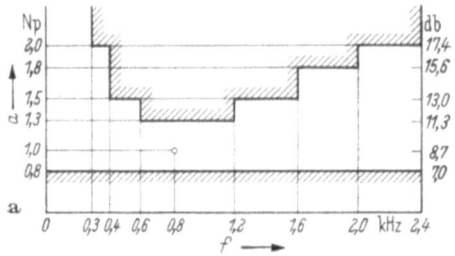

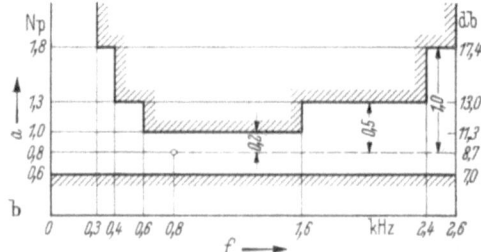

Figs. 31a and b. Limits of overall equivalent in point-to-point traffic (CCITT-Recommendations, 1938).

a) International two-wire circuit; b) International four-wire circuit

mission. The latest recommendations provide for a range of from 300 Hz to 3400 Hz (bandwidth 3.1 kHz) (Fig. 32.)

In carrier telephony over postal lines all efforts are concentrated on transmitting a large number of communication channels between two terminal stations. To permit optimum utilization of the transmission lines, the individual SSB speech channels are arranged side by side. Guard bands of from 600 Hz to 900 Hz must be provided between the individual channels to allow for their segregation in the receivers with filtering media not exceeding a reasonable cost level. This results in a 4 kHz "frequency allocation scheme" for postal lines.

The first carrier communication systems for operation over power distribution networks were designed long before the first CCITT recommendations were published. The engineering problems differed greatly from those encountered in the communication systems of postal administrations. The transmission task consisted in providing merely one telephone circuit for each line section in a mesh whose sections could not be decoupled from each other. There was no need whatsoever for multichannel operation. Terminal equipment transmitting the carrier and the two sidebands up to a range of $\pm$ 2.4 kHz proved to afford the best solution, both economically and technically.

In countries where the construction of an extensive carrier communication system was started at an early time, the 5 kHz allocation scheme has become firmly established because of the large number of existing 5 kHz terminals. The tendency now is to regard this merely as a historical

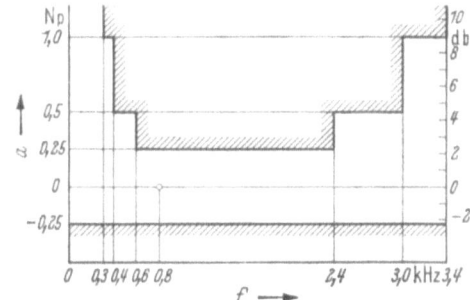

Fig. 32. Limits of overall equivalent in point-to-point traffic (CCITT recommendations, 1946)

relict, outdated by the new 4 kHz CCITT scheme. This attitude is not justified. As a matter of fact, two different problems are involved. The merits of the allocation schemes must be evaluated independently and in the light of their capability of meeting the following two sets of requirements:

a) For the purpose of postal communications as many voice channels as possible have to be accommodated on a line connecting two terminal stations. The same frequencies can be used again and again on all sections of the postal network (Fig. 33a), because the individual lines are completely isolated from each other.

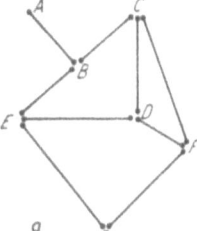 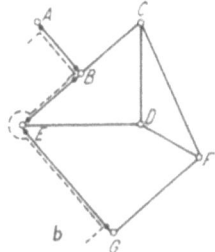

Figs. 33a and b. Carrier communication networks.

a) Postal communication network; b) Power line network

b) The power distribution network (Fig. 33b) on the other hand, constitutes a mesh which is metallically interconnected at the nodal points (station bus bars). Communication depends only on individual carrier circuits between the various stations, in no case on channel groups as dense as those found in postal networks.

5 Podszeck, Carrier Communication, 4th Ed.

Power companies are of the opinion that the improvement in speech transmission, envisaged with the extension of the voice band to 3400 Hz cannot be brought to bear in their operating area and that the wider frequency bands would unnecessarily contribute to the existing frequency shortage. To this must be added the fact that a receiver designed for a 4 kHz band will pick up more noise than a 2.5 kHz receiver. Adoption of the 1938 CCITT recommendation has therefore been limited to the two-wire circuits (Fig. 31 a). These minimum requirements, however, should be strictly observed. Otherwise, quality of speech transmission in a single line section would deteriorate to a point where tandem connection of several sections would result in very poor long-haul communication. The same reasoning applies also to SSB terminals used for telephony and the simultaneous transmission of telemetering information. These "multi-purpose terminals" would comply with the new CCITT recommendation if the speech band were extended up to 3400 Hz. This, however, is not done for the reasons discussed above. They are in compliance with the older CCITT recommendations if the speech band ranges up to 2400 Hz.

For teleprinter and telemetering applications the frequency bands need not be nearly as wide as for telephony. The earlier telemetering functions of the power companies, such as telemetering on the pulse frequency principle or the remote control of selector switches were based on a telegraph transmission method providing for speeds lower than those of conventional teleprinters. A telegraph speed of 50 bauds[1] could therefore be assumed as an upper limit for signal transmission. Satisfactory transmission of telegraph signals was ensured at a bandwidth corresponding to 1.6 times the telegraph speed, in our case 1.6×50 Hz $= 80$ Hz.

After the conversion of the telemetering terminals from electro-mechanical to electronic components, telemetering methods permitting higher telegraph speeds are used to an ever increasing extent, which require broader transmission channels. For this purpose even multiples of the 120 Hz spacing, recommended by the CCITT for 80 Hz channels, are used, i. e. 240 Hz or 480 Hz. These frequency allocation schemes, taken over from voice frequency telegraphy, provide for a much closer channel spacing than the allocation schemes established for accommodating telephone channels.

In a band sufficient for transmitting speech frequencies it is therefore possible to accommodate a number of telegraph channels for telemetering purposes, e. g. 18 channels of the 120 Hz scheme in a 2.5 kHz band or 9 channels of the 240 Hz scheme or 4 channels of the 480 Hz scheme. For

---

[1] Unit of telegraph speed [8].

speeds higher than 200 bauds, as is normal practice in data transmission, telegraph channels for 600, 1200 or 2400 bauds are used.

If the data channels are not dedicated channels but are transmitted instead of speech over telephone connections, which are established by dialing if and when they are required, "modems" (mo/dulator – dem/odulator) are used. These equipments are the links between the data terminal equipments, which supply dc signals, and the transmission path, over which these signals are transmitted in the form of VF signals.

The factor 1.6, used for an approximative calculation of the bandwidth, decreases with high telegraph speeds. Nevertheless, the frequency bands required for telegraphy will still be wide enough to prevent a voice channel from being loaded with more than one high speed telegraph channel.

In carrier communication on overhead power lines, the signal delay on the line is of little consequence compared with the delay introduced by transients and other characteristics of the filters installed in the terminal equipment. The build-up time of a filter is a function of the bandwidth and, by way of approximation, may be said to be inversely proportional to the bandwidth. It amounts to some 10 ms in a filter designed for a passband range of 100 Hz. Aside from the bandwidth, filter delay depends also on the number of filter sections, and it is in many cases larger than the build-up time. The filter delay approximates a value equalling $0.7 \frac{n}{2} \frac{1}{\varDelta f} s$, where n is the number of filter sections and $\varDelta f$ the bandwidth. In the filters designed for a 120 Hz allocation scheme and employed in teleprinter transmission systems, the sum of build-up time and delay time is in the order of approximately 30 ms. This total delay is unacceptable in a high-speed network protection system. A channel of the 480 Hz allocation scheme will be required for such applications, if this sum is to be kept down to a value not exceeding some 8 to 10 ms. Transmission bands even wider than these must frequently be employed in practice to provide the necessary transmission speed (see page 127).

In power line communication networks it is only the 5 kHz or the 8 kHz allocation schemes which are of interest as well as the 2.5 kHz and 4 kHz assignments derived therefrom, the former for DSB and the latter for SSB transmissions. Since adequate decoupling of individual line sections of a power network is not possible, great care must be taken in assigning the frequency bands of single-channel telephone terminals in such a manner that the same carrier frequency will not appear again within a wide radius around the transmitting terminal. Frequency assignment is here a problem which must be dealt with again every time a new carrier circuit is added or the configuration of the mesh is changed.

The 5 kHz allocation scheme has been established for pure telephone channels (single-purpose equipment). Superimposed signaling channels (multi-purpose equipment) cannot be added at a later time in case of double-sideband operation, because the frequency band provided by this allocation scheme is not sufficiently wide. If only a single signaling channel were required in addition to the telephone channel, it would be possible to accommodate it between the carrier frequency and the lower edge frequency of the voice band, without exceeding the 5 kHz limit. Since more than one channel is required in most of all cases, a separate 5 kHz channel, loaded with a group of narrow signaling channels, is used for each direction of transmission. The transmission tasks quite naturally led to this practice, because at the beginning of the development nobody thought of anything else but speech transmission. Signaling channels were required only at a later time and on a much smaller scale. SSB transmission with single-purpose equipment remains the preferred choice even today, if speech transmission and signal transmission are to be treated as completely separate functions and a maximum of flexibility is to be retained for their accomplishment.

Another factor may also have influenced the decision of power companies to use single-purpose terminals. It is the desirability of having completely independent equipment for speech and signal transmission. This segregation permits a certain amount of control over power system conditions to be retained if the one or the other type of equipment is out of service for periodic maintenance or repair. On the other hand, where only a small number of signaling channels is required, this separation of functions means higher initial costs.

Multi-purpose DSB terminals operate within an 8 kHz allocation scheme. They cannot be readily integrated into an existing 5 kHz assignment. Therefore, they can be employed only in such networks as include but a few, or no terminals, designed for another allocation scheme (Fig. 34).

It has often been argued that multi-purpose equipment is a money-saving investment because the signaling channels can be added at a later time, as the requirement arises. While this is true, one must not overlook the fact that an allocation scheme with a wider frequency spacing will then be required and this implies tying oneself down right from the beginning to a scheme which provides fewer voice channels than could be obtained with single-purpose equipment. A really satisfactory utilization of the available frequency range can be secured only if the requirement for signaling channels can be brought into agreement with the capabilities of multi-purpose equipment, considering all angles such as number of channels, direction of transmission, and geographical location. A long-term forecast of such requirements can hardly be made, however, and when it finally becomes clear that the advantage of progressively added signaling channels was gained only at the expense of frequency space it will be too late for remedial action.

The rigid coupling of telemetering and telephone channels, involved in the use of multi-purpose equipment, can therefore hardly be reconciled with the necessity of frequency conservation. Most telemetering channels are operated in one direction only, while the narrow frequency band available for the opposite direction remains unassigned and serves no

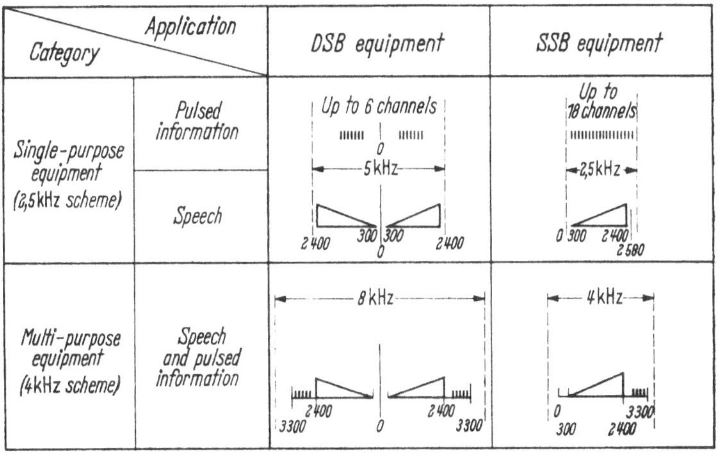

Category \ Application		DSB equipment	SSB equipment
Single-purpose equipment (2,5kHz scheme)	Pulsed information	Up to 6 channels	Up to 18 channels
	Speech		
Multi-purpose equipment (4kHz scheme)	Speech and pulsed information		

Fig. 34. Relationship between category of equipment, application, and frequency assignment

purpose whatsoever. On the other hand, if telephone channels are combined with teleprinter channels, or with line protection channels, the situation is entirely different, because these signaling channels are always worked in both directions.

SSB equipment has been designed for the 5 kHz and the 4 kHz allocation schemes to achieve a true saving in frequency space. The 5 kHz scheme suggested the idea of using one frequency position either for telephony and a number of superimposed signaling channels, or for transmitting two contiguous telephone channels in the same direction ("two-channel telephone terminal"). The 4 kHz band, however, can be loaded only with one telephone channel and a few superimposed signaling channels, since the band is too narrow for accommodating two telephone channels. Arranging a 4 kHz multi-purpose SSB terminal in the 5 kHz allocation scheme results in a loss of 1 kHz per frequency position. It will then be better to use a two-channel terminal equipped with a single telephone channel and to load the second voice band with a correspondingly large group of signaling channels.

In double-sideband transmission it is normal practice to state the center frequency for denoting the frequency position within the allocation scheme. In the 5 kHz scheme, for instance, transmission over the 100 kHz channel means transmission in a band ranging from 97.5 kHz to 102.5 kHz. In the case of SSB transmissions within this allocation scheme, the nominal frequencies will coincide with the junction between two SSB channels. An agreement must therefore be reached on whether the nominal frequency of the SSB circuit refers to the channel adjoining the nominal frequency from above or from below. At the same time, new frequency positions are created for SSB transmissions, whose nominal frequencies lie between the nominal frequencies of DSB equipment, such as the 97.5 kHz and 102.5 kHz positions in a 5 kHz allocation scheme. To have figures which can be more easily remembered, the nominal frequencies of the 5 kHz scheme are simply retained and a + sign or a − sign is added as an index to distinguish the upper from the lower position, 100 kHz − and 100 kHz + in our example.

For denoting the positions in an 8 kHz or a 4 kHz allocation scheme it would be unpractical to use the center frequency. Instead, the upper or the lower edge frequencies of the positions are used to designate a channel. A simpler and frequently adopted method is the numbering of frequency positions.

If SSB terminals are designed so as to afford an optional spacing between the go-channel and the return-channel, the send and receive frequencies of a station can be fitted into a DSB allocation scheme so that inter-channel crosstalk is kept within permissible limits. The unassigned band of the original scheme may then be used for superimposed signaling channels in the same carrier terminal, if it is a multi-purpose equipment, or for carrier terminals at another location. A two-channel telephone terminal occupies two 5 kHz positions, one for two contiguous 2.5 kHz channels in the outgoing direction, the other for two such channels in the incoming direction.

A rigid spacing between the two channels of a telephone circuit is likely to involve additional difficulties in frequency assignment. This applies in particular to a fixed zero spacing, when both channels are worked immediately side-by-side. It would appear that a 5 kHz or an 8 kHz position, accommodating only one channel in one direction with DSB transmission, could now be used in SSB transmission for contiguous channel working in both directions. In other words, a frequency position previously assigned to a single receiver would now be used to operate, at the same location, a transmitter and a receiver each on one half of the original band. It will be necessary to check whether this transmitter interfers with other carrier terminals operating on neighboring frequency positions. It can be said that SSB equipment for contiguous band working will provide a generally acceptable and convenient solution to frequency assignment problems only in those cases where earlier-type equipments do not exist or where they can be replaced by new ones irrespective of the costs.

5.3 Transmitter Power and Range

Power engineers tend to use the transmitter power as a yardstick for assessing the quality of a carrier communication terminal operating over power lines. The axiom that the send power should be as large as possible in order to obtain a satisfactory signal-to-noise ratio at the receiving end is certainly justified. Since the noise level on power lines is much higher than that on postal lines, power line carrier terminals must of necessity be designed for an appreciably higher send power than the carrier terminals used by postal administrations.

The output of a carrier frequency transmitter intended for transmission over power lines has in many countries been limited by administrative regulations in an effort to avoid interference with private radio sets or industrial radio receivers.

The endeavor to maintain a satisfactory balance between efficiency and cost of a transmitter is another factor which contributes to a limitation of the send power. In this connection it is worth mentioning that the rise in cost is anything but proportional to the rise in send power. Referring the output power P_{tr} of a transmitter to the internationally accepted standard power $P_0 = 1$ mW, the send level p_{tr} can be derived from the equation

$$p_{tr} = 10 \log \frac{P_{tr}}{P_0} \, \mathrm{db} \,.$$

From the relationship between power and level (Fig. 35) it will be seen that a deviation of the send power from its nominal value has,

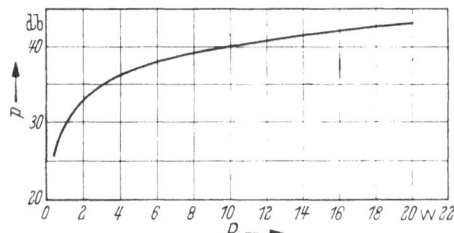

Fig. 35. Send level versus send power

within certain limits, no appreciable effect on the send level. To give an example, a nominal power of 10 W corresponds to a send level of 40 db, and a nominal power of 8 W to a send level of 39 db. A rating of 10 W has come to be generally accepted as a standard value, to be measured at the output terminals of the equipment, no matter whether the bandwidth is 8 kHz, 5 kHz, 4 kHz, or 2.5 kHz. In exceptional cases, a send power of up to 300 W is used in some overseas countries. The necessary power amplifiers are here not part of the carrier communication terminal. They are supplied as additional items to the carrier terminal, where required.

The reasons for increasing the send power as far as administrative regulations and considerations of economy permit are detailed in the following:

a) The noise level on a power line depends on the line construction and on the maintenance efforts expended. A line which may be sufficiently good for power transmission may be rather poor for the purpose of carrier communication, in particular under adverse weather conditions. Nevertheless, it should be capable of satisfying the requirements of message communication to the greatest possible extent, without the necessity of costly improvements on the power lines. An insulator disc, a transformer bushing, or a ground lead which, while showing minor imperfections, is still good enough for power transmission, will raise the HF noise level far beyond the normal. The same is true of a power line conductor whose diameter was initially chosen for a lower operating voltage and which is now required to carry a higher voltage. In these and in similar cases one should like to have a reserve in signal-to-noise ratio so that not every deterioration in transmission conditions impairs the reliability of the communication system. Even though the send level may be raised up to the permissible limit to counter adverse conditions, it will in any case be more profitable to eliminate the sources of noise.

b) The interference level, which may appear in any position of the frequency scheme and which originates in high-power radio transmitters whose radiation is picked up by the power line, even at a large distance from the transmitter, may be so high that it is not drowned in the general noise level. It is in fact an increment to the inherent noise and the sum of these two noise components is decisive for the signal-to-noise ratio of the carrier communication circuit. This is one more reason for a high useful level and, consequently, for a high send level.

c) A high send power is also required where a number of channels has to be transmitted with the aid of a common send amplifier. Multichannel telemetering terminals, or a multi-purpose equipment for simultaneous speech and superimposed signaling channel transmission, may here be mentioned as examples. This distribution of the total available send power must not result in too low a level for each individual channel and in a correspondingly low signal-to-noise ratio at the receiving end.

It is not the output power alone which counts. The transmission methods (SSB and DSB depending on the depth of modulation; FM depending on the frequency swing) must also be taken into account. Their influence on the quality of transmission is by far greater than that of the send power. Amplitude modulation in connection with DSB or SSB terminals is the preferred operating option for transmission over power lines.

Each transmission terminal has a "frequency position" assigned within the allocation scheme, a fixed frequency band, which determines the passband range of the line filters of the carrier terminals. In message transmission with DSB terminals the frequency position is occupied by a lower sideband, the carrier frequency, and an upper sideband. The upper and the lower sidebands contain the same message. The available transmitter power is distributed among these three components. In the case of SSB terminals, the carrier is suppressed and only one of the two sidebands is transmitted. The result is a by fare more efficient utilization of the send power. This advantage must be paid for in terms of higher initial costs. Using frequency-modulated carrier terminals, a frequency position is taken up by the carrier and the first pair of sidebands, as in DSB transmission (Appendix 9.6).

When a number of messages are transmitted with a common equipment, as is the case in multi-channel terminals for telemetry or telephony, and in multi-purpose equipment for simultaneous speech and telemetering signal transmission, the demands placed on the send amplifier are higher than those which must be met in single-channel transmission. While in the latter case the available send power can be fully used for transmitting a single message and an occasional overloading of the amplifier is not very critical, it is of great importance in multi-channel operation that the output of the send amplifier be suitably distributed among the various channels, and that the amplifier be never driven to the overload limit. Overloading must here be avoided because the resulting sum and difference frequencies would appear as noise voltages in the various channels.

From what has been said above it becomes obvious that the send power P_{tr} is limited in the first line by the maximum output P_{max} which the send amplifier is able to deliver. It depends on the transmission method, on the other hand, how high a send power can be obtained from an amplifier without overdriving it and how the send power is distributed among the sub-channels within the frequency band to be transmitted.

The transmission range r of a terminal is a function of the send level p_{tr} and the noise level p_n. It is further determined by the minimum signal-to-noise ratio Δp_n, required at the input of the receiver to ensure reliable operation:

$$r = p_{tr} - (p_n + \Delta p_n)$$

The noise level picked up by a receiver depends on the bandwidth. Since the frequency band for SSB transmissions is only half as wide as that for DSB transmission, the noise level at the input of a SSB receiver is smaller by $10 \log 2 = 3$ db than that at the input of a DSB receiver.

An analysis of these relationships (9.7) reveals the influence of the send power and its distribution among the sub-channels, as well as the influence of the various transmission methods on the transmission range. A minimum signal-to-noise ratio of 26 db has been established both for signaling channels and for telephone channels.

5.4 Radiation

When electrical energy is transmitted over physical lines, a circular magnetic field is formed around the conductors and an electric field between the two conductors. The energy is propagated perpendicularly to these two fields in the dielectric between go and return conductors [6]. In a lossless, homogeneous and infinitely long line there is no essential energy radiation at the side. This radiation is limited to the line field, which decreases at either side of the line the closer go and return lines are run. Theory indicates that the field strength decreases with the square of the distance.

Power lines, however, are not constructed same as the line used for theoretical calculations. Vertical components must be assumed because of the line sag between towers and because of the transposition towers. In particular phase-to-ground coupling of a carrier frequency system at one end of a line having no corresponding return conductor constitutes a Hertzian dipole which radiates energy. Phase-to-phase coupling, too, is not completely balanced to ground. All vertical leads act as omnidirectional antennas.

This is why, although the "short-range field" decreases with the square of the distance, some conductor elements cause at certain points a "far reaching field" which decreases only linearly with the distance. It should be noted, however, that this field decreases also with the frequency.

Thus, power lines themselves may be a source of interference to radio receivers, in particular to commercial radio sets. This is due to the high-frequency noise voltages initiated in the power distribution network, such as corona or switch operations. In addition to this, a power line carrier communication circuit may disturb radio receivers located in the vicinity of the power line. On the other hand, a power line acts as an antenna picks up energy from the HF field of external transmitters, and it is possible that this energy disturbs operation in the carrier communication circuits of the power line network.

As far as river and coastal navigation services and air traffic control are concerned, radio reception is protected by the postal administrations ensuring a carefully planned distribution of the frequency positions to radio stations and to power line carrier communication systems. In the case of commercial radio receivers, which can be tuned to a wide range of frequency bands and whose location with relation to the power line is not defined, it will be necessary to clarify the question of up to what distance from the power line radiation is still effective, and what simple measures offer themselves to eliminate interference. Radio reception in close proximity to the power line is greatly impaired, if not entirely impossible, in view of the interference introduced by power transmission. At a certain distance from the power line no interference whatsoever will be noticed, neither from power transmission, nor from carrier signal transmission. What must be determined is the range between these two limits.

The amount of carrier energy radiated by the line is rather low. This is why carrier communication systems cannot be counted among radio transmission systems, from a technical point of view. Radiation remains within reasonable limits because the separation between the go- and the return leg of a carrier communication circuit, i. e. the separation between the power conductors is small in relation to the wavelengths used for carrier communication. The amount of HF energy radiated by the power line depends firstly on the carrier output of the transmitter (not in excess of 10 W) and secondly, as far as radiation near the coupling points is concerned, on the type of line coupling. At some distance from the coupling point the non-coupled conductors of the power system take over the function of the return leg, and the amount of energy radiated in a transverse direction is then approximately equal for phase-to-phase and for phase-to-ground coupled carrier systems. The attenuation of the send power along the transmission path increases with the distance from the carrier transmitter and this involves a proportional decrease in vertical radiation from the power line.

Considerable work has been in progress in various countries to determine the boundaries of the field produced by radiation. Experiments with a tuned frame antenna, erected at a height of 6 to 10 feet above ground, or on roof tops at some 50 to 65 feet above ground [31, 32], revealed the fact that the lines of force are arranged approximately parallel to the ground at distances of 100 yards to 500 yards from the power line. An investigation into the behavior of the field at right angles to the power line, using a vehicle-mounted field strength analyzer of the type employed chiefly in the broadcasting range from 150 kHz upwards, confirmed in all cases the result which was to be expected on the basis of previous calculations: the field strength decreases with the distance x from the power line at a ratio closely approximating $1/x^2$ (Fig. 36).

At a distance of not more than 10 to 20 yards from the power line, the largest values of the field strength were found to range between 10 mV/yard and 100 mV/yard. The peak values measured approached some 300 mV/yard. From these measuring results, the field strength curve can be calculated with $1/x^2$ taken as dropping rate.

Distance x (yards)	100	200	500	1000	2000	3000
Field strength in μV/yd	1000 to 3000	250 to 750	40 to 120	10 to 30	2.8 to 7.5	1 to 5

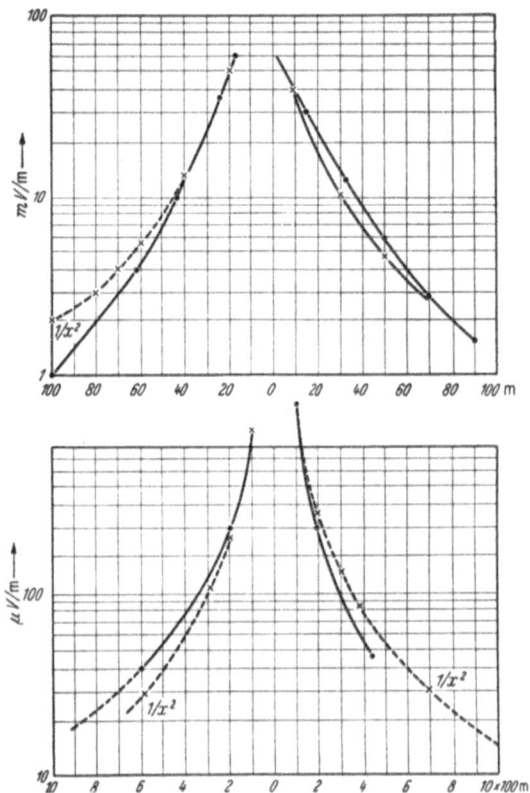

Fig. 36. Field strength of a 230 kHz carrier versus distance from power line (in the vicinity of a carrier transmitter having an output power of 10 W)

In view of the static noise level and the residual noise in DSB operation (radio broadcasting), a received field strength of less than 100 μV/yard can hardly ever be used for practical purposes within the frequency range of long-wave transmitters. For radio reception, current practice is to base calculations a supply field strength of 1 mV/yard and on a signal-to-noise ratio of 1 : 100. It follows that an interference field strength of 10 μV/yard is no longer regarded as a disturbance. In practice, the interference is no longer of importance at a distance of more than 1000 yards, measured perpendicularly to the power line. In the cases so far reported, trouble occurred almost invariably at distances of less than 600 yards.

The grounded steel structures of the power lines tend to decrease the field produced by the carrier communication circuit. Besides, the clearance between ground and conductors is greater at these points than on

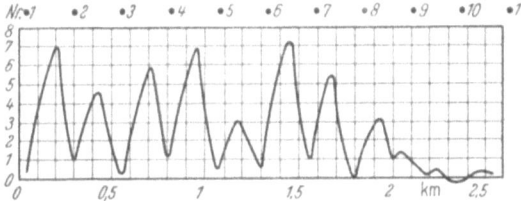

Fig. 37. Field strength produced by a carrier frequency of 230 kHz along a power line at a distance of approximately 5 yards from the conductors, send power 10 W

the points of maximum line sag. As a result, the field strength curve shows dips at the points coinciding with towers (Fig. 37).

As an easy way to determine, during planning, which values the field strength emanating from the carrier frequency system must not surpass in the neighborhood of power lines it may be assumed that only comparatively narrow noise frequency bands are involved, so that the probability of a disturbance is low. A compilation of theoretical deliberations and measurements results in planning values which differ for the neighborhood of the lines and the neighborhood of the stations (Fig. 38).

A deterioration of radio reception due to radiation of carrier energy is more likely to occur in the vicinity of low-voltage power distribution networks, since these lines frequently cross densely populated rural

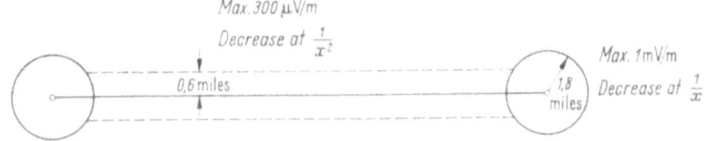

Fig. 38. Planning values for the maximum permissible radiation

areas. It is in particular the continuous transmissions, as in telemetering applications, which are likely to impair radio reception. On the other hand, unintentional listening to service messages is also undesirable as is intentional monitoring. The latter is however possible only at carrier frequencies higher than 150 kHz without previous modifications to the radio sets whose operating range normally begins approximately at this frequency.

If the carrier and both sidebands are transmitted, monitoring of messages can be prevented by means of "speech inverters" which translate the voice band into its inverted position before modulation. At the receiving end, the voice band is re-inverted after demodulation, so that the conversing parties can talk over the line as over any normal circuit, while monitoring along the carrier section is not readily possible. Although a carrier may be added in the monitoring equipment, the listener will have some difficulty in understanding the message because an interference tone of a relatively high level is produced by the added carrier in conjunction with the transmitted carrier.

Special "speech scramblers" are available. Unauthorized monitoring by third parties is here possible only if the auxiliary frequency used for speech scrambling is known, and this frequency can be varied at certain intervals in accordance with a prearranged schedule. Little use is made of these devices because of their high cost. If there is a need for frequent transmission of confidential service messages, provision of a carrier channel for teleprinter operation will be a simpler solution.

Single-sideband transmission with suppressed carrier presents difficulties to listening-in, which are inherent in the transmission method. If the carrier is fully suppressed, it must be added again to permit monitoring. A radio receiver would here be required which can be caused to oscillate at the suppressed carrier frequency, in other words, a regenerative receiver. This type of equipment, however, is no longer in regular production.

Monitoring of an SSB transmission with suppressed carrier by means of a regenerative receiver is anything but convenient, because the frequency of such receivers is not sufficiently constant. Specific investigations have shown that continuous retuning is necessary if the listener wants to understand the message in its context and not only fractions of it.

If it is not a case of intentional monitoring but of an undesirable disturbance of radio reception, the situation can in most cases be remedied by relocationg the antenna or by replacing the radio set.

6. Carrier Communication Networks

The term "carrier communication network" refers to the entirety of the message channels of an extensive carrier communication system. Depending on the type of traffic, a further classification is made into "telephone networks" consisting of a number of "telephone districts", "teleprinter networks", and "telemetering networks". A "carrier frequency district" includes transmitters and receivers tuned to the same frequency.

When engineering a carrier communication system, the frequency assignment problem turns out to be of fundamental importance right from the outset. The rapidly growing requirements for new communication channels have to be reconciled somehow with the comparatively small frequency range which is made available to power companies. In view of the insufficient decoupling of the individual sections of the power mesh and the precautions necessary to avoid interference with radio broadcasts, the assignment of frequencies is a very delicate part of the planning work. This applies in particular to densely populated, highly industrialized countries.

Telephone networks and telemetering networks have different functions. As long as single-purpose equipment is used, the problems involved can be treated separately and the boundaries are clearly defined. These boundaries become blurred where multi-purpose terminals are employed. The feeder lines to the carrier network must be designed with a view to preventing an impairment of the safety and reliability of the entire communication system by inadequate input and output circuits.

6.1 Frequency Allocation Scheme

The frequency shortage in extensive networks is chiefly due to the fact that a carrier frequency used on one section of the power network will appear also in remote sections with a level high enough to cause response of the receivers of other communication systems, if these receivers happen to be tuned to the same carrier frequency. A carrier frequency can therefore not be used a second time unless the geographical distance is such that the level of these disturbing remnants is drowned in the general noise level. Undesirable propagation of the carrier frequency outside its legitimate boundaries takes place partly by way of the metallic path across the traps, specifically intended to confine the message path, and partly by inductive and capacitive coupling with conductors lying parallel to the conductors actually coupled. A really satisfactory precondition for using the same carrier frequency again in another nearby section of the power network could be created by the provision of crosstalk traps. Because of their high cost, however, such traps are only rarely employed.

Inadequate by-pass circuits are another source of undesirable propagation of carrier frequencies within the power network. This trouble source can be easily eliminated (see page 41). A newly inserted junction line, merely intended for load compensation between two nodal points of the power network, may also lead to difficulties in frequency assignment within the existing communication network, even though this junction line may never be required for carrier operation.

The allocation of free frequency positions within the overall scheme is governed by considerations which sometimes fail to satisfy the (more or less) justified interests of the various system operators to the extent desired by individuals. The risk that new equipment, to be integrated into a system, will disturb existing equipment must be borne by the equipment manufacturers. A thorough knowledge of the technical data of the existing carrier terminals and line equipment is therefore indispensable. The performance data of interest are send level, sensitivity, and selectivity of the carrier terminals, blocking impedance and bandwidth of the wave traps, as well as passband attenuation and selectivity of the coupling media.

If the available frequency spectrum is already rather crowded, the selection of the carrier frequencies to be used for a new system is mostly made in close cooperation between the manufacturer of the carrier equipment and the power company. This approach takes also account of the retuning operations which had to be carried out on the existing facilities in connection with previous network modifications. In this way a joint survey of the latest status of the frequency assignment is prepared before the carrier frequencies for new transmission terminals are assigned. Poor judgment in the selection of suitable carrier frequencies may involve the necessity of extensive retuning at a later date, and this may turn out to be a costly affair. Where optimum utilization of the spectrum is to be achieved, increasing congestion naturally implies higher risks. Frequency allocation plans are based on the assumption that the transmitter of a communication circuit will, as a matter of principle, operate with its maximum output while the sensitivity of the associated receiver is, in the case of short lines, reduced by means of a fixed attenuator in its input circuit. Reducing the send power and utilizing the full sensitivity of the receiver would be an unpractical alternative, since this would result in an unnecessary decrease in the signal-to-noise ratio. If adjacent frequency positions, or the same position, are to be assigned twice within the allocation scheme, a level diagram (Appendix 9.5) must be worked out to verify that, considering the possibility of undesired HF energy dissemination in the power network, the carrier energy has fallen off to a level sufficiently low to permit the same position or adjacent positions to be used for other purposes.

Frequency allocation schemes are preferably presented in a self-explanatory manner (Fig. 39) in connection with a carrier circuit schematic. Instead of this diagrammatic presentation, an index card file with specially shaped cards may be used. Where coupling between networks of different operating voltages must be expected because their conductors are run in parallel over a large distance, frequency assignment must be coordinated by making corresponding entries in the diagram or by using file cards of different color. Recent attempts to computerize frequency planning have as yet not produced a generally applicable method.

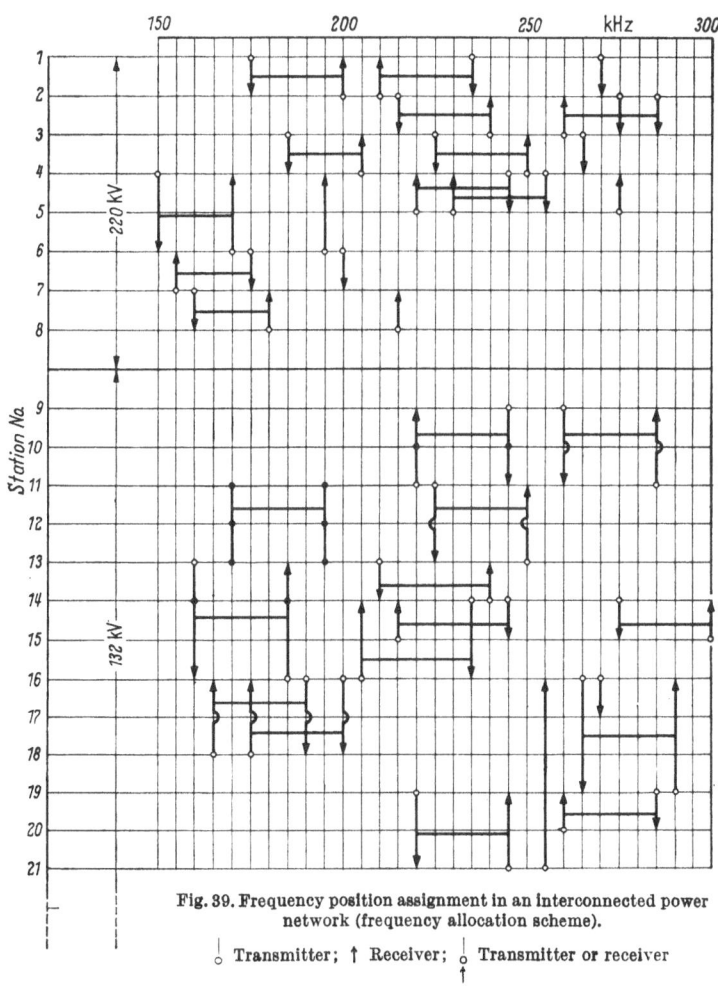

Fig. 39. Frequency position assignment in an interconnected power network (frequency allocation scheme).

○ Transmitter; ↑ Receiver; Transmitter or receiver

6.2 Telephone Networks

At a time when amplitude-modulated DSB terminals were used almost exclusively, attempts were made to avert the imminent frequency shortage at first by combining as many carrier terminal stations as possible in one district. Such efforts are, however, hampered by two limiting factors. Firstly, the transmission range of a communication channel, limited by the send level and the sensitivity of the receiver, must not be exceeded. Secondly, the traffic density, growing proportionally with the number of the send stations operated over the same channel, must not be allowed to reach a point where busy conditions occur too frequently.

Telephony implies in each case a type of "both-way communication", in other words, each party must be able to listen and to talk. A telephone circuit is therefore invariably worked in both directions of transmission. The most satisfactory option is full-duplex traffic, where both ends are able to talk simultaneously. The carrier terminals used for speech transmission occupy two transmission channels in the allocation scheme, one for the outgoing and one for the incoming direction. The transmitting branch and the receiving branch of a telephone carrier terminal are therefore tuned to two different carrier frequencies. As a result, one carrier frequency pair must be set aside for each telephone district, and only one call can be made at a time.

Some carrier terminals, operating on a "half-duplex" basis, require only one frequency position in the allocation scheme. Transmitter and receiver of this type of equipment are tuned to the same frequency, and the two directions of transmission for talking and listening are selected by change-over from transmission to reception. This change-over need not be a manual operation, it may as well be voice-controlled. Since this mode of operation tends to complicate the design of the exchange equipment in a selective-dialing telephone network, this type of equipment has not found widespread application and will therefore not be discussed in more detail.

There are various options for combining telephone carrier terminals in a telephone district and consequently, different modes of operation (Fig. 40). In a telephone district designed for "point-to-point traffic", the transmitter works with a permanently assigned carrier frequency to the correspondingly tuned receiver of the distant station. A saving in carrier frequency pairs (or telephone terminals) can be achieved by combining more than two telephone stations in one district, instead of using each power line for a single point-to-point district. The number of the stations which can be combined in this way is determined by the previously mentioned limiting factors (transmission range and traffic density). In the case of "radial traffic", more than two stations can communicate with each other in such a manner that a message is sent from a central station to one sub-station or to a number of sub-stations on one carrier frequency, while the second carrier frequency serves for transmissions from the sub-stations to the central station. Communication between the sub-stations is here not possible.

The "frequency interchange principle" is adopted if, in a district including more than two stations, connections are to be established between any two stations. In the rest condition, the receivers of all stations combined in a district are tuned to the same carrier frequency, anytime ready for reception. When one of the stations has a message to transmit, its transmitter must be switched over to this carrier frequency. Lifting

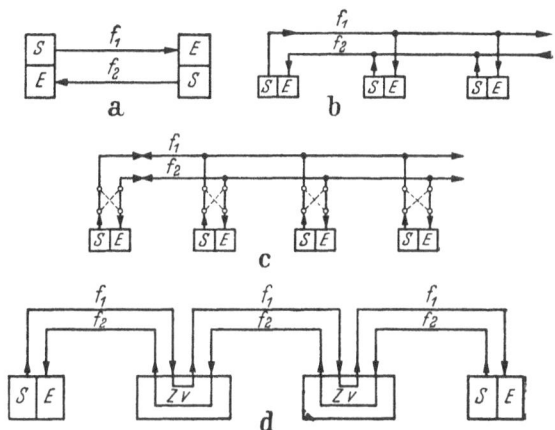

Fig. 40 a–d. Traffic options in telephone districts.

a) Point-to-point traffic between two stations; b) Radial traffic between several stations; c) Frequency interchange traffic between several stations; d) Way-station traffic between several stations; T Transmitter; R Receiver; WR Waystation repeater

the handset, the calling party causes the transmitter of the carrier terminal to be switched to the same carrier frequency to which the receiver was previously tuned, while the receiver is switched over to the carrier frequency assigned to the distant transmitters.

The "party-line" or "way-station technique" has been developed with a view to further extend the limit set to the possible number of telephone stations in a district by the range of the transmission terminals. The carrier terminals at both ends of the line (district) transmit their carrier frequencies continuously with no frequency interchange taking place. The way-stations are equipped with an intermediate carrier frequency repeater for each direction of transmission. In these (way-station) repeaters calls from and to the local stations are added and dropped respectively. The VF currents are switched under the control of the call number and in dependence of whether the call is to proceed over the one or the other repeater output to the corresponding carrier channel.

The communication options differ for the various modes of operation (Fig. 41). Radial district networks are infrequent because of an undue limitation of communication options.

Waystation traffic is ideal where bandwidth is at a premium, because a larger number of stations can be combined in one telephone district due to the fact that one of the previously mentioned limitations (transmission range) can be overcome by using way-station repeaters. On the

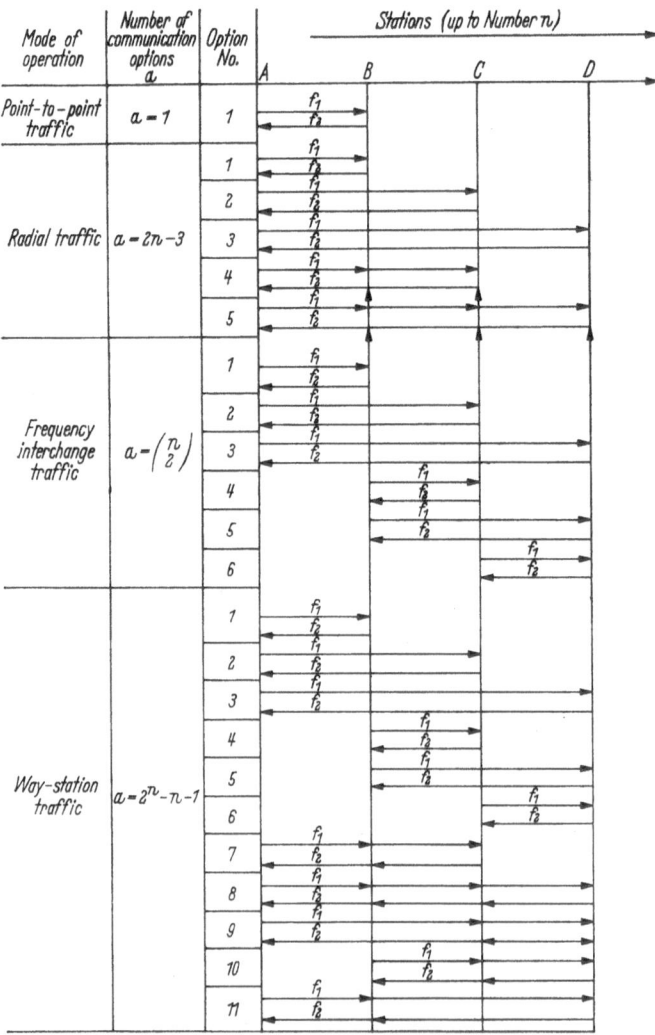

Fig. 41. Communication options with the various modes of operation

other hand, the second limitation (traffic density) takes on an increasing importance. Telephone districts operating on the frequency interchange principle comprise three, at the most four telephone stations, because the limit of the range will then be reached in most cases.

While the carrier frequencies of the HF transmitters and receivers remain unchanged in point-to-point, radial, and way-station traffic, any telephone station may be either transmitter or receiver for each of the two carrier frequencies assigned to a district when the frequency interchange principle is employed. Thus, the frequency of a transmitter is no longer tied-up with a fixed location. This tends to aggravate the situation, as far as frequency allocation is concerned. It may be somewhat relieved, without restricting the communicating options within a district, by deactivating the frequency interchange between transmitter and receiver in one of the stations. This "deactivated frequency interchange" makes it necessary to introduce either "call number dependent frequency interchange" or a "frequency dependent interchange" in the remaining stations of the district, to avoid a sacrifice in communication options. Frequency interchange in dependence of the call number is possible by introducing a discriminating prefix or a call number memory. Frequency interchange in dependence of the frequency is accomplished by arranging the filters for the two frequencies ahead of the receivers so that they are in parallel during the rest condition. Depending on the incoming frequency, one of the two filters is disconnected from the receiver and switched over to the transmitter.

Deactivation of frequency interchange in a station is resorted to in those cases where either send or receive frequencies only are to be assigned to contiguous positions of an allocation scheme in an effort to avoid crosstalk. If this is done in more than one station of a district, the communication options will be correspondingly restricted.

All these operating options, with the exception of point-to-point traffic, have been tailored to the specific communication needs of power companies. In the past, power companies were merely confronted with the problem of establishing only one lightly loaded telephone circuit along the right-of-way of a power transmission system. This simple transmission task tends to facilitate efforts directed towards conservation of frequencies and equipment. The operating options described developed into what may be termed a particularity of power line carrier telephone communication and have found widespread use. They have never been adopted by other communication services, neither in postal nor in railroad networks. Selective dialing on party lines might, with some reservations, be likened to the way-station traffic or to the frequency interchange traffic of the power companies.

As long as only one telephone circuit is required and normal availability standards can be maintained, waystation traffic will offer the most favorable solution to the frequency conservation problem, because more carrier terminals can be combined in one district than would be possible with frequency interchange. But the larger the number of cascaded repeater sections operating on the same frequency pair, the more difficult will it be to find two frequency positions which will avoid conflict with other channels over so long a total span. To do without a revision in the existing frequency assignment, conversion of the carrier frequency in one of the waystation repeaters, i. e. a sacrifice in spectrum space, might become inevitable.

For several years there has been a development away from this practice. The growing tendency to adopt selective dialing in an effort to keep pace with the increasing traffic volume generated by the continued expansion and ramification of intermeshed networks compels power companies to abandon the old methods in favor of those employed in postal networks. The trend now is to interconnect the various stations of a power network so as to form a "network group" which permits each station to contact any other station by direct dialing. Provisions are to be incorporated which minimize the chance of finding any one line section busy. The carrier frequency sections between the nodal points of the network are operated on a four-wire basis, as in postal networks. This implies point-to-point traffic. A single telephone circuit between nodal points is frequently found to be inadequate for the traffic volume to be handled and light channel groups, comprising about 3 to 6 channels, have to be provided.

Although 3 or 6 of these single-circuit telephone terminals could of course be operated in parallel over the same line section, reasons of economy dictate the use of multi-channel equipment. The costs of each individual telephone circuit decrease if the basic costs of the racks, cabinets, power packs, metering circuits, and carrier transmitters and receivers can be prorated to a larger number of circuits. This fact suggested the idea of adopting the carrier terminals originally designed for use in postal networks.

This trend has been followed in several countries for quite some time although mostly for different reasons. Originally, the intention was not so much to provide light channel groups for establishing intermeshed networks, but to cover the comparatively low equipment requirements for the narrow sector of power line carrier communication from a larger production volume, in other words, the emphasis was on avoiding costly special development wherever practicable.

In the course of time it became apparent, however, that a close adaptation to the particular needs of the communication services of power companies could not be avoided. This is based on the following reasoning:

a) Postal transmission systems are designed for a voice band ranging from 300 Hz to 3400 Hz, i. e. for a 4 kHz frequency allocation scheme. Instead of the 2.5 kHz position, each voice channel occupies a 4 kHz band without a corresponding profit. The provision of a larger number of voice channels of this bandwidth tends to unnecessarily aggravate the frequency shortage.

b) Postal transmission systems are intended for a rigid 4 kHz scheme without any provision for flexibility in the reassignment of frequencies. Only in countries using the 4 kHz scheme also for power line carrier terminals (with identical frequency positions), can these systems be readily fitted into the allocation scheme and only under these conditions can the channel units of postal systems be used for all frequency positions up to the limits of the permissible range.

c) The channels are arranged side-by-side for the incoming and the outgoing direction, so that two frequency bands are required whose width increases with the number of channels. In the case of a 6-channel group, this amounts to two frequency bands, each with 6 times 4 kHz = 24 kHz. So wide a frequency position can be made available in an existing carrier communication network only with graet difficulty and at considerable expenses. Convenient integration is possible only in newly planned communication networks, or in power networks requiring only a limited number of carrier circuits.

d) Carrier transmission terminals taking up so large a bandwidth involve increased difficulties also as far as the line equipment, i. e. wave traps and coupling circuits, is concerned. Adequate compensation of the attenuation distortion on a transmission line section, to achieve equally satisfactory conditions for all channels, becomes all the more critical the wider the frequency band is that has to be transmitted. Two or three pilot channels are therefore used to distribute equalization functions to several portions of the frequency band. Transmission terminals affording two or three telephone circuits are quite frequent, whereas a 6-channel telephone terminal approaches the limit of the currently used multi-channel terminals. Admittedly, these channel numbers agree very well with the requirements so far encountered, and the frequency space which they occupy can be made available without major difficulties.

e) In many cases the problem is not one of point-to-point traffic between the nodal points of a mesh, but of transmission to a common terminal station, which may be a load dispatcher's office. This deviation from postal transmission tasks entails further complications.

f) The output of the send amplifier is in most cases insufficient for operating postal transmission terminals over power lines, if a signal-to-noise ratio is to be maintained which ensures reliable operation also over the long distances involved. In such cases it will therefore be necessary to interpose a separate power amplifier between the carrier transmitter and the line because the postal practice of using intermediate repeaters just at those places where they are needed cannot possibly be applied to power line carrier communication.

In the case of carrier transmission over the bundled conductors of 380 kV power networks (see page 44) it might perhaps be possible to find ways and means to eliminate the difficulties a) to d) connected with frequency assignment. The postal transmission terminals can then be adopted and the line groups required in these networks may comprise as many as 24 channels.

In the usual power networks, however, the aforementioned limitations to the use of multi-channel terminals and the difficulties involved by the adoption of postal systems must be expected to continue unless four-pole traps (Appendix 9.4) are resorted to which ensure adequate decoupling of a line from the remaining sections of the power network. Experiments have been conducted again and again with multi-channel telephone terminals, in particular with a view to finding an answer to the question whether power lines can be used as a transmission medium for the public communication services, where densely packed voice channel groups are the rule. This problem was tackled in the Soviet Union as early as in 1932 and today it is again of considerable significance to several overseas countries. The construction of radio relay connections in these countries is in some cases rendered difficult by unfavorable topographical conditions and their operation, i. e. power supply and maintenance, may be too much of a problem. Attempts are therefore made, when new and extensive power networks are being erected, to accommodate also some groups of voice frequency telephone channels intended for the public telephone service.

In large carrier communication networks special demands are placed on the individual telephone district, on the through-connection between the districts, and on the transition to voice frequency telephone installations. These demands are related to the quality of transmission. Frequency assignment, transmission range within a carrier district, and the remaining previously discussed carrier communication problems are not directly affected.

Telephone circuits were originally intended merely to establish a means of communication between two stations in a power network. When several not too distant stations were later combined so as to form a telephone district, speech was again transmitted over only one channel for each direction of transmission. When the power networks were extended so as to cover also remote regions, it became necessary to connect several carrier frequency districts in tandem, i. e. to transmit speech over a number of series-connected channels. Even sub-stations separated by distances exceeding the transmission range of the carrier terminals are now able to communicate with each other. This was a parallel development in several countries. When the circuits originally intended for district traffic had to be switched through at the transit stations by the time interconnected system operation was started, designers were faced with the problem of long-haul communication.

Telephone circuits, even if each individual circuit operates satisfactorily, cannot be simply connected in tandem in the hope that the total connection would, of necessity, also permit satisfactory speech transmission. If the quality of speech is poor in a connection extending over several carrier districts, or including a transition from HF to VF sections, it might well be that such a failure must be attributed to this erroneous notion.

In a telephone connection the frequencies of the 300 Hz to 3400 Hz voice band are attenuated to a different degree, even if speech is transmitted in its original frequency position. In the case of an unloaded cable line, for instance, the higher frequencies are attenuated more than the lower frequencies. This dependence of the attenuation on the frequency, the frequency response of a telephone circuit (Fig. 42), is compensated by "equalizers" so that as equal an attenuation as possible is obtained for all frequencies of the voice band. However, it is only the "frequency-

Fig. 42. Frequency response within the voice band (linear distortion) for one direction of transmission.

a Without equalizer; b With equalizer

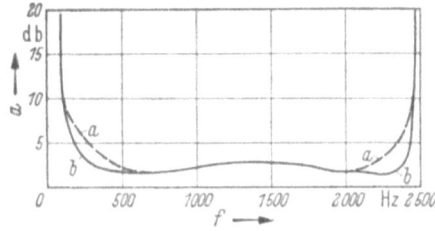

dependent attenuation distortion" (linear distortion) which can be reduced in this way, but not the "amplitude-dependent non-linear distortion". Later equalization of non-linear distortion is not possible.

Thus, the quality of speech transmission is decisively influenced by the attenuation (which is in most cases given for a mean frequency of 800 Hz), the linear distortion, the non-linear distortion, and the noise level, all measured in the voice frequency position. These four components may vary within certain limits before this variation is felt to be annoying by the conversing parties at both ends of a communication circuit. In short-haul communication within a single district, the values of these factors may be less favorable than those which are permissible in the individual sections of a long-haul communication circuit comprising a large number of tandem-connected telephone districts. To prevent the sum of attenuation, distortion, and noise level values of the individual ectio ns of a long-haul circuit from exceeding the permissible limit, the values measured in the individual districts must be smaller than those which can be tolerated in a telephone district intended exclusively for short-haul communication.

If a voice channel has been translated in one section of the transmission path from its natural position into the HF position, a distinction must be made, in stating attenuation and distortion values, between the values applicable to the HF section, and those pertaining to the voice frequency channel plus HF section. The quality of speech is determined exclusively by the values measured in the voice frequency channel.

If an observation of the level conditions in a carrier telephone terminal starts with the microphone of the talking party, the VF send level measured at the input of the carrier terminal is found to be a value to which the various circuit units of the carrier terminal, up to the modulating stage, are permanently adjusted in such a manner that an optimum modulation index is obtained for the HF carrier. Variations within certain limits of the VF input level may occur where VF stations located at different distances are enabled to obtain the carrier circuit by way of a private automatic exchange. Such variations are permissible although they involve a change in the depth of modulation. This applies, in analogy, to the receiving end. The amplitude of the speech current delivered by the converter stage depends on the depth of modulation at the transmitting end. The level across the VF output terminals of the carrier equipment is raised by the VF amplifier in the receiving branch to a point sufficiently high to reach the desired station with an adequate useful level, irrespective of the differences in line attenuation which must be expected when the call is routed through a private automatic exchange at the receiving end.

The send level in the HF position is a function of the output power delivered by the HF send amplifier. The attenuation on the power line varies with weather conditions and with the switching conditions of the power network so that the useful HF level at the receiver is likely to drop in an unpredictable manner. In extreme cases, the HF signal-to-noise signal ratio must not drop below 26 db (see page 74). It should be noted that the HF noise level may also increase with a change in the weather. The volume control circuit connected ahead of the HF receiver ensures a constant HF input level for the converter stage.

It has been found convenient to interconnet carrier frequency districts in a way different from that adopted for VF telephone systems, where a single wire pair is available between subscriber and exchange. A voice frequency switchboard interconnects the two wires arriving from the calling subscriber with the two wires leading to the called subscriber. This is a "two-wire switchboard" for either manual two-wire through-switching or automatic two-wire through-dialing, depending on whether the switching system is of the manual or automatic type. With circuits connected in this way, the communication currents are transmitted in both directions in the same original frequency position.

In carrier telephone communication a separate transmission channel is available for each direction of transmission. A circuit of this type may be regarded as having a wire pair for one direction, and another wire pair for the opposite direction. This amounts to regular four-wire traffic. The transition from two-wire to "four-wire traffic" is effected within the carrier terminal by means of a hybrid (see page 114). This hybrid involves an attenuation of the speech currents because the differential transformer distributes the power equally among communication path and line balancing network. Each telephone district terminating hybrid causes an attenuation of $10 \log 2 = 3$ db.

With a transition from four-wire to two-wire operation, this hybrid reduces that portion of the energy which flows from the receiver directly to the transmitter, to prevent self-oscillation (singing) of the circuit. Two adjacent telephone districts should not be through-connected by way of a two-wire switchboard, because singing must here be prevented by extra padding. This can be avoided by using a four-wire switchboard, i. e. by switching the currents supplied by the receiver to the transmitter of the outgoing channel in the voice frequency position. The transition from four-wire to two-wire traffic, required for serving local stations, is avoided in the inter-connection of carrier districts either by "four-wire manual through-connection" or by "four-wire through-dialing", depending on wether the switching equipment is designed for manual switching or for direct subscriber-to-subscriber dialing.

Cases where circuits for more than ten carrier terminals have to be switched in a nodal point of the network are infrequent. The number of connections to a four-wire switchboard will therefore always remain small. Normally, all carrier terminals connected to a four-wire exchange have the same status, i. e. no distinction is made as to privileged and non-privileged stations.

In a manual switching system an attendant may handle calls in the order of their importance to power system operation. He may even separate existing connections in favor of urgent calls. In direct dialing systems, all subscribers are connected to the carrier terminal by way of an automatic exchange. Privileged subscribers are assigned a direct connection to the carrier terminal and a separate call number within the telephone district. This direct connection affords the possibility of entering a call in progress by operating the entering key. It is possible to enter a connection established with a subscriber of the same carrier terminal, or a four-wire through-connection ("local entering"). In a frequency interchange district, a connection may be entered which has been made between two other carrier terminals of the same HF district ("district entering"). If a distant station in a third HF district is to be afforded the possibility of entering a connection established between two other HF districts at a four-wire through-dialing station, a special entering call number may be introduced. In this way access to busy districts other than the one to which the calling station belongs can be obtained.

For establishing connections on a four-wire basis it proved to be convenient to provide equipment both for direct dialing and for manual switching. Special functions, such as breaking in on a conversation, night service during unattended periods, etc. can best be accomplished if both switching options are available in parallel. These special duties include the through-connection of long-distance calls of immediate importance to system operation as well as the supervision of telephone traffic, to discourage stations of another network from setting up connections to other stations within their network by way of through-dialing points in neighboring networks.

In discussing the tandem connection of a number of circuit links, only those viewpoints have so far been dealt with as are of significance to the quality of speech transmission. Some additional conditions must, however, be met to ensure also proper transmission of the dial pulses.

The run-down time of the dial switch in a telephone set has been standardized for the digit having the longest pulse train, i. e. for a time of 1 ± 0.1 s. The pulse-to-interval ratio, also fixed, amounts to $0.40 : 0.60$. The relays and selector switches in the dial exchanges have been adapted to these standard values. What is meant by a distortion of dial pulses is the same prolongation and shortening of pulses and intervals as occurs in telegraph signal transmission [8]. "Pulse timers" are provided at the end of each transmission section to regenerate the dial pulses so that they are again within the specified tolerance limits before they are relayed over the next following section.

Power line carrier telephone terminals have either their own built-in relay unit, in particular if they are designed for frequency interchange traffic, or they depend on external switching facilities to perform the necessary relaying functions. An outgoing dial train is transmitted with the built-in dial pulse send relay, while an incoming selection is accepted by a dial pulse receive relay. If the carrier terminal is equipped with a relay unit, a DC loop is normally required for connecting the operator's telephone set. As in the case of a connection to an automatic exchange, the dial pulses are transmitted to the carrier terminal by the dial switch contact interrupting the DC loop, fed from the carrier terminal, in the cadence of the pulse train. The dial pulses arriving from the distant station serve to set up a route to the desired subscriber which is called by AC ringing pulses, sent over the same loop by means of a vibrator converter.

There are, however, cases where the selected route involves a transition from carrier frequency circuits to voice frequency circuits which are terminated by transformers. This applies to communication lines influenced by power lines. To ensure the necessary protection of personnel and equipment, such lines must be terminated by insulating transformers. Another type of transformer (matching transformer), is required for adaptation to the characteristic impedance of loaded cables. Additional facilities must be provided in these cases to ensure transmission of the dial signals over the isolated link.

Where similar conditions are encountered in the long-distance traffic over postal cables, the dial signals are transmitted in the VF position to permit the same transmission channel to be used both for speech and dial signals. Considering the short local loops, this method would be too expensive in power system operation. It will here be sufficient to use an inexpensive and mains-independent AC source which does not depend on a complex auxiliary current source. "Inductive dialing" is frequently employed as the most simple means of dial pulse transmission over lines terminated by insulating transformers. The dial pulses are here obtained from a DC source, originally provided for feeding the local relay circuits and passed on to a specially designed transformer, the "pulse repeater". This pulse repeater transmits a dial pulse in the form of two associated current pulses of opposite polarity. Although the number of transformers which can be bridged with inductive dialing is small, this method will in most cases prove sufficient for power system operation.

Repeaters inserted in a VF feeder line must be by-passed if the frequency of the dial signals falls outside the voice band which the repeater will allow to pass.

The number of the digits of a call number may be rather high in systems designed for through-dialing, in particular if suffix dialing is provided in connection with automatic exchanges. Storages built into the automatic switching unit proved to be a useful means for reducing the number of digits. Since several digits of a call number must first be dialed before an indication is received as to whether the path to the wanted remote district is available, this storage circuit may be used as well for automatically continuing the selection procedure as soon as a previously busy section has been cleared. In other words, the calling party need not start dialing again. A special "busy indicator" system may be provided to report from one to another nodal point which of the intermediate telephone districts are occupied, and a long-distance connection will then be set up only when the calling party is assured that all circuits up to the desired station are actually available. These busy indicator systems operate over separate channels similar to switch position indicator systems. They avoid unnecessary seizure of line sections due to unsuccessful calls.

The call numbers of all directly connected stations in a carrier communication network operated on the direct dialing principle are assigned in accordance with a call number scheme. The routes to the wanted station, which may extend over several telephone districts, are determined by different call numbers. The principles on which a call number scheme is based differ with the design of the automatic exchange system. The call number scheme covers only the direct connections to the carrier network. Two numbers, for instance, are assigned to each carrier terminal providing access to two electrically interlocked stations. It must not be confused with the directory, which includes all station call numbers of the entire communication system. This directory completes the listing by stating the suffix numbers of those stations which are tied in by way of private automatic exchanges connected to the carrier terminals.

6.3 Telemetering and Remote Control Networks

Unlike telephone communication systems, telemetering and remote control systems cannot be used for exchanging all sorts of information. Their range of activity is therefore mostly limited to the power network of a single company. Long-distance transmission over a number of cascaded sections is here a rare occurrence. Although measurands for network control are occasionally transmitted between remote stations by way of long-distance telemetering channels, long-haul traffic will hardly ever be required for any other supervisory control applications.

The demands imposed on signaling channels as regards transmission reliability are necessarily higher than those imposed on telephone channels [36]. Mutilated dial signals result merely in a wrong connection or in an unsuccessful call. Unsatisfactory quality of speech may be largely compensated by efforts on the part of the conversing parties to adapt themselves to unfavorable transmission conditions. Undependable pulse transmission in supervisory control systems, on the other hand, may result in a falsified picture of the operating conditions within the network or in extreme cases, prevent operation altogether. While it is true that supervisory control systems are designed in such a way that interference pulses or missing pulses will not result in a wrong indication or in an unwanted control action, falsified telemetering signals are likely to mislead an analog recorder into simulating load peaks or load valleys which actually do not exist.

In assessing the dependability of transmission channels the various duties arising in telemetering and remote control applications cannot be accorded the same rank. One might arrive at wrong conclusions if this problem is not clearly understood. Thus, for instance, existing telemetering equipment, designed to meet only relaxed dependability standards to save costs, may some day be reassigned, in connection with network modifications, and then be used for the transmission of remote control or line protection signals. In such cases the carrier terminal would have to be redesigned for increased dependability or replaced by an equipment meeting higher standards.

Telemetering and remote control systems do not in each case depend on two-way circuits (Fig. 1). Thus, only a single message channel is required for transmitting measurements to recording instruments, or for transmitting counter readings. The carrier terminals are either pure transmitters or pure receivers and not a combined send/receive equipment, as in telephony. If, in exceptional cases, a channel must be worked in both directions for transmitting brief signals on a half-duplex basis, e. g. for supervisory control, the same type of carrier terminal may be used as is adopted for telemetering. Switching over from transmitter to receiver and vice versa at the same carrier frequency is effected by the remote control equipment.

Whereas multi-channel telephone terminals for use on power lines are operated only with a few telephone circuits because a dense grouping of telephone channels meets with technical and economic difficulties, wide use has been made of multi-channel telemetering and remote control terminals. Since a bandwitdh of 80 Hz (see page 66) is sufficient for telegraph speeds up to 50 bauds, a total of 18 such channels can be accommodated in a 2.5 kHz position, otherwise taken up by a single speech channel. In view of the existing frequency shortage it is desirable that each channel be used only for continuous transmissions, e. g. for telemetering, rather than for widely spaced and very brief signals of minor importance.

Such signals may be transmitted in sequential order over a common channel, with entering rights assigned on a priority basis, if necessary.

Continuous transmissions of, say, metered data, need not necessarily take up one channel for each variable, as is the case in "frequency-multiplex" systems. A number of variables may be sampled consecutively on a time basis and then transmitted over a common channel. Such a "time-division multiplex" terminal, which may cope with up to 20 variables in each one-second cycle works with synchronized distributors (multiplexer and demultiplexer) at the sending and at the receiving stations. This is a fully-electronic equipment, containing no moving parts subject to mechanical wear.

The telegraph speed in a channel operated on a time-division multiplex basis is higher than the speed of a single-message channel. A wider frequency band must therefore be made available. An equally large number of variables, transmitted on a frequency-multiplex basis, requires a still broader frequency band. Thus, for instance, the transmission of 20 variables would require a frequency band of $20 \times 120 \, \text{Hz} = 2400 \, \text{Hz}$. The same number can be accommodated in a $360 \, \text{Hz}$ channel with time-division multiplexing, if a one-second cycle time is acceptable. If this cycle time is decreased, a correspondingly increased bandwidth is required. During the intervals between the individual sampling instants the variable is stored in memory circuits.

Time-division multiplex depends on a reasonable ratio of the cycle of the electronic distributor and the rate of change of the signals currents to be transmitted. The sampling period assigned to each variable within a full cycle must be sufficiently long to permit proper interpretation of the pulse code, which represents an information or the magnitude of the variable.

If high-speed electronic coders are employed, a time-division multiplex terminal may operate at a low cycle speed. A sufficiently long sampling period to permit proper interpretation of a digital or of an analog value is then all that is required. In some applications, where the slower electro-mechanical coders are employed, the time-division multiplex terminals must operate with shorter cycle times to ensure a sampling rate which is high enough to avoid unpermissiple signal dostortion. Such applications are the remote control by means of selectors [4, 9] and the transmission of pulse trains of unspecified length and long signal duration (pulse rate metering, counting).

Time-division multiplex terminals, as a means of saving frequency space in the carrier communication networks of utility companies, are employed only in those cases where an existing narrow frequency band is to be loaded with more channels than would be possible with frequency multiplexing. The choice is therefore limited to comparatively slow multiplexers which can be operated over narrow channels. Time-division multiplex terminals having a high cycle speed are unlikely to play a significant role in present carrier communication systems because of their high cost and the increased bandwidth requirements.

Whereas time-division and frequency-division multiplexing may be regarded as pure transmission problems, other methods of arranging transmission processes on a time basis come within the scope of the local automatic switching units. This results in the alternate use of carrier terminals. A variable may be transmitted, for instance, during the idle periods of a telephone circuit. A more frequently adopted arrangement provides for the interruption of speech for fractions of a second to transmit network protection signals and similar information. Frequency and equipment conservation is the objective of any alternate operation of multi-purpose systems. An interruption of speech, however, is tolerable only if the signal to be conveyed is very short. Otherwise, transmission is possible only in telephone districts of low traffic density. To avoid a complication of circuitry as a result of this chronological arrangement of transmissions of different priority rating, alternate operation of carrier terminals is restricted to only a few easily controlled combinations.

On the other hand, multi-purpose equipment designed for the simultaneous transmission of speech and telemetering signals has found a wide range of application. In some countries it has been firmly established as a standard equipment within a correspondingly arranged frequency allocation scheme. This has been done although, in some cases, it will not allow ideal adaptation to the transmission duties of a carrier communication network, and without regard to the fact that requirements as to maximum dependability and frequency economy cannot always be easily met (see page 70). The installation of frequency-interchange districts is obviously impossible. While frequency interchange is readily feasible with single-purpose telephone terminals, it is incompatible with the need for continuous transmission between a transmitter and a receiver, a need which arises in the large majority of all telemetering and remote control applications. The use of multi-purpose equipment frequently involves the necessity of providing a larger number of telephone districts than would be required if single-purpose equipment and frequency interchange were adopted. The result is invariably a waste of frequency positions.

A clear cut separation between telephony and telemetering applications will greatly benefit the flexibility required in the planning of extensive communication networks, since both networks can then be best adapted to specific needs. A different type of equipment is therefore required, i. e. single-purpose equipment for telephony and similar equipment for telemetering and remote control, the two versions being different, constructionally and electrically. The question whether completely independent networks or the use of multi-purpose equipment will finally provide the most economical and frequency-saving solution depends on the requirements to be met in each specific case.

The decision whether a carrier network is to be established with single-purpose or with multi-purpose transmission terminals is of fundamental importance as far as frequency assignment is concerned. Using single-purpose equipment, one may load the frequency positions either with telephone channels or groups of signaling channels, leaving oneself free up to the last minute to make a final decision. In the case of multi-purpose terminals, however, one is tied down from the outset to terminal points being identical for the telephone and the signaling channels. The 4 kHz allocation scheme affords only a small number of superimposed signaling channels. Although up to ten channels of 80 Hz bandwidth can be theoretically accommodated between the upper cut-off frequency of speech (2400 Hz) and the cut-off frequency of the full voice channel (4000 Hz), only 6 signaling channels can be actually used because standardized VF channels of the 120 Hz scheme are not available for frequencies above the 3300 Hz cut-off point, and because these higher frequencies can no longer be transmitted over loaded feeder lines.

This limited number of signaling channels available in the 4 kHz allocation scheme is admittedly sufficient for practical applications, provided the channels are transmitted, at a slight increase in bandwidth, on a time-division multiplex basis. However, the number of variables to be transmitted would then have to be rather large, to justify the installed first cost of a time-division multiplex system, which is invariably higher than that of a frequency-multiplex system.

In telemetering systems broad transmission channels for high telegraph speeds are permanently through-connected (dedicated channels), e. g. in the case of a process computer being used in a load distribution center or in the case of time-division multiplex with high cycle speed. In conventional remote data processing, on the other hand, modems are frequently used which are connected into the public switched telephone network. In the latter case either voice or data are transmitted because the transmission band of a modem is broad and bacause its position cannot be altered in view of the standardization of the versions. A connection is established by dialing a number. After handshaking the operators at both ends switch over to data transmission. Remote switching in the called station from the station of the calling operator is also possible by dialing a number. Since several seconds are required for connection set-up prior to actual data transmission, this mode of operation is not suitable for the transmission of telemetry signals.

To transmit messages over great distances, several carrier frequency districts must be connected in tandem. Corona noise and delay distortions in the individual sections are responsible for the fact that only 600 baud channels can be used with a bandwidth of 2.1 kHz (2.5 kHz allocation scheme) and 1200 baud channels with a bandwidth of 3.1 kHz (4 kHz allocation scheme). From 2 to 4 sections can be cascaded without the need for special measures. Particularly careful equalization and special modulation methods permit modulation rates up to 2400 bauds in the frequency bands of the 4 kHz allocation scheme.

There is, of course, little sense in transmitting at high speeds over an excessively disturbed carrier frequency channel. Each individual error makes it necessary to repeat the entire signal sequence and in this way so much time can be lost that the same effect can be achieved with a slower equipment which is less susceptible to noise, i. e. with a less expensive equipment.

6.4 Transition to Communication Feeder Lines

Power lines supplying large cities are mostly terminated at the outskirts of a city in a transformer sub-station, where the carrier communication terminals are also set up. The administration buildings of the power companies are normally located some miles away in the center of the city so that telephone cables or radio circuits must be used to connect them. Even if telephone communication over these local feeder lines is satisfactory and the long-distance traffic over the power line communication circuit is also flawless, interconnection between the telephone sets in the administration building and remote stations of the carrier network by way of the feeder lines and the switchboard in the sub-station will result in a satisfactory long-haul communication path only if the previously mentioned conditions regarding attenuation, distortion, and noise level are satisfied.

Carrier telephone terminals for use in power line communication networks are designed to provide the same voice frequency levels at both ends of the channels as proved to be convenient in postal carrier communication networks. If a level of -17 db across $600 \, \Omega$ is available at the input of the HF transmitter (modulator), a level of $+8.7$ db will be delivered by the HF receiver as the demodulated output. If the remote VF station is to operate at a level established as satisfactory for communication over the postal network, the voice[1] must be impressed upon the line at a level of 0 db, and received at a level of -6 db. At relaxed speech quality standards the linear distortion on the feeder lines may amount to 4 db. It must under no circumstances exceed 1.7 db, if the circuit concerned forms a link within a longhaul communication circuit. These two figures represent the maximum permissible deviation of the attenuation at the two cut-off frequencies of the voice band (300 Hz and 2400 Hz), related to the standard 800-Hz tone. If the attenuation values at the cut-off frequencies are both higher or both lower than the standard value at 800 Hz the larger of the two deviations must not exceed 4 db and 1.7 db, respectively. If the deviations have different sign, their difference must not exceed the aforementioned figures. These standards define the limits of linear distortion permissible on the feeder line. The amount of linear distortion can be visually displayed by means of a level recorder (Fig. 43).

Open-wire telephone lines strung on the structures of a power transmission system and influenced by the power line are frequently more difficult to connect to a carrier communication system than non-in-

---

[1] Since the talker output itself varies greatly, a standard 800 Hz test tone is used as a basis for measurements.

fluenced telephone lines. Because of the larger distances between the telephone stations and the switchboard, the higher attenuation introduced by the protective devices of the power line, and the higher noise level, considerable efforts would have to be expended in many cases to provide in the entire voice frequency network the preconditions for through-connection to the carrier network.

In an extensive voice frequency network with many subscriber stations and switching facilities it will be better, for technical and economical reasons, to superimpose a carrier frequency circuit on the

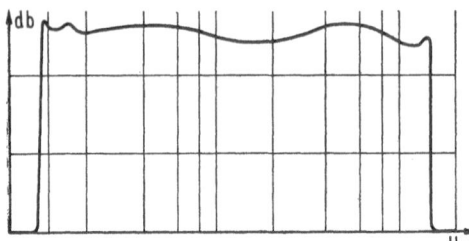

Fig. 43. Level characteristic

voice frequency transmitted over the influenced open-wire line. In this way, independence from all switching facilities within the voice frequency network is achieved. Coupling to exposed open-wire telephone lines is much less of a problem than coupling to a power line, because neither the power voltage nor the power current must be taken into consideration. Simple lowpass filter networks ahead of the transition to the voice frequency telephone system may be used as traps for the carrier frequency message, and highpass filter networks ahead of the transition to the carrier network as traps for the voice frequency band. The components of these filter networks have to be rated for an insulation level equalling that of the exposed telephone line, i. e. for some 10 kV.

The switchboards of the voice frequency communication system are by-passed by two lowpass filters and one highpass filter. Voice frequency telephone equipment in way-stations connected in parallel to the line need not normally be provided with lowpass filters to avoid carrier losses, because the stray inductance of the insulating transformers will moslty suffice to take care of this problem.

There is nothing in particular to be said in connection with frequency allocation, since the exposed open-wire telephone lines are strung on the structures of medium-voltage lines and no carrier terminals are coupled (otherwise, a VF station could be coupled directly to the power line by means of a carrier terminal). If, in exceptional cases, a power line and a telephone line, suspended from the same towers, are both used for carrier communication, a common frequency plan must be worked out.

Where open-wire telephone lines are used as feeders, the carrier terminals are the same as those employed for transmission over power lines. This is done for the sake of uniformity in equipment, if it can be economically justified, or in anticipation of a later change-over of the feeder line to a higher operating voltage, to have the same carrier terminals available for operation over the new power line.

By-pass circuits have also been designed for the direct transition from power networks to exposed open-wire telephone lines so that a telephone station in the medium-voltage network can be connected directly to a carrier frequency telephone district of the high voltage network, without a switchboard being interposed at the point of transition from a medium-voltage network to a high-voltage network.

No problems will arise where only one signaling channel has to be provided on feeder lines connecting to a carrier frequency telemetering network. In most cases, however, an entire group of VF channels has to be transmitted over a feeder line of considerable length. The transmission equipment is then divided into a HF section installed in the substation located outside the residential area, and a VF section set up in the downtown administration building. What has been said about telephone communication with respect to attenuation, distortion, and noise level [8] applies essentially also to VF transmission sections used for supervisory control purposes.

6.5 Carrier Current Relaying for Network Protection

Protective relaying must be provided for all power transmission systems to automatically disconnect faulted line sections within fractions of a second. Protective relays are installed at each end of the line to supervise the system and to detect overcurrents, ground faults, and any change in the direction of current flow. In the less complicated cases it is sufficient to trip the power circuit breaker only by these locally detected criteria [7]. Such applications are "inverse time overcurrent protection" (tripping times delayed in dependence on the magnitude of the overcurrent) or "distance protection" (tripping times stepped in dependence on the distance). It is only the comparison systems which require a communication channel between the two ends of a power line, since the results of the measurements are compared before the circuit breakers are tripped simultaneously at both ends. Protective relaying systems depending on carrier channels come within the scope of telemetry. They are used chiefly for spanning great distances, in particular in the 220 kV and higher tension networks.

The power networks become ever more ramified and growing requirements are placed on the power carrying capacity of the lines. This entails a commensurate increase in the short circuit current which must be brought under control in the event of a line failure. The trend in power circuit breaker design is therefore predominantly in the direction of obtaining increased interrupting capacities and higher tripping speeds. The detection of the location and nature of a fault and its disconnection from the system are to be accomplished at ever increasing speeds. In the case of a temporary fault, the selectively disconnected section of the mesh must be reconnected as quickly as possible for the sake of service continuity. This implies the use of "reclosing-type relays".

A selective line protection system particularly suited to meeting these requirements in power system operation is distance protection with reclosers [38]. To reduce the tripping time of the normal distance relays and to ensure parallel operation of the circuit breakers at both ends of the line, a connection is here required for transmitting the protective signals between the two stations which terminate the line section. The most natural solution is the use of carrier circuits on the power line to be protected, i. e. some sort of carrier coupling for high-speed distance protection. The distance relay being closer to the point of fault will be quicker in transmitting the tripping signal to the associated circuit breaker than the remote relay. Carrier frequency coupling is to ensure high-speed tripping also for the remote end of the line.

The comparison systems, i. e. all selective line protection systems which, employing signaling channels, can be classified into the category of telemetering systems, may be grouped in the following manner:

a) Current comparison (differential protection);
b) Phase comparison;
c) Selective line protection operating on the
    c1) transfer tripping principle,
    c2) transfer blocking principle;
d) Intertripping, depending on
    d1) change of relay setting,
    d2) fault detection (starting relays),
    d3) tripping signal only.

Intertripping, as far as it depends merely on the transmitted tripping signal (d3), is of importance not only for line protection, but also for generator protection, if the generator and the circuit breaker are not accommodated in the same station. Carrier equipment, similar to the carrier terminals for network protection, is here required to transmit the tripping signal between the protective relay of the generator and the circuit breaker over an intermediate power line section.

The carrier terminals are here required to render dependable service just at a time when normal attenuation conditions on the transmission path, which is the power line, are disturbed by a short circuit or by a ground fault, and when excessively high HF disturbances are caused by arcing so that impulse-type noise of high amplitude, large bandwidth, and irregular pulse sequence is superimposed on the normal noise level. Of special interest are therefore the operating conditions on power lines under these abnormal circumstances which differ greatly from regular conditions (section 4), so that particularly exacting requirements must be placed on carrier terminals intended for protective signal transmission.

A short circuit in the power network may be single-phase to ground or multi-phase to ground and, for experimental purposes, it may be assumed to be almost a metallic connection. Experience has shown that multi-phase short circuits and multi-phase ground faults are rather infrequent. Nevertheless, the selective line protection system must be designed so as to react also to these faults. The characteristic of the incremental attenuation caused by such faults may be represented for phase-to-ground and phase-to-phase coupling in dependence on the distance between the point of fault and the ends of the line. Unlike a metallic connection, the arc has not an impedance of zero Ω, but one of 5Ω to 20Ω. As a result, the incremental attenuation is actually smaller. At any rate, it will rise considerably in the end zones of the line. It is much lower for phase-to-phase coupling than for phase-to-ground coupling. This is one of the reasons why phase-to-ground coupling is avoided in network protection systems. Phase-to-ground coupling would, most unfortunately, involve a high incremental attenuation just in that case where carrier relaying is intended to reduce the tripping time of distance relays, i. e. the dependability of tripping signal transmission would be impaired if the failure occurs near the end of a line. Guiding values have been determined also in the phase-to-phase coupling case for the incremental attenuation caused by a short circuit.

Ground fault	Incremental attenuation approximately
Single-phase, non-coupled phase	1.7 db
Single-phase, one of the coupled phases	4.5 db
Two-phase, one of the coupled phases affected	3.5 db
Two-phase, both of the coupled phases affected	13 db

Fig. 44. Incremental attenuation with a short-circuit current of 3500 A at a distance of 875 yards from one end of the line

An incremental attenuation of some 20 db and still higher values have been measured for a two-phase to ground fault with phase-to-ground coupling and the coupled conductor affected.

A failure on a live power line causes not only a higher attenuation but also an increased noise level. Impulse-type noise of high-amplitude, the individual noise bursts differing in pulse density and length [39], is superimposed on the normal, uniform corona noise.

Ground faults and short circuits are frequently caused by lightning. Each partial discharge induces an interference pulse. Two partial discharges were found to be the average, while 40 discharges have so far been found to be the maximum. Discharges occur at irregular intervals of 8 ms to 400 ms.

Flashovers may result in short-circuit arcs [19]. A strong noise pulse appears when the arc is established and a weaker one when it breaks. Without carrier coupling, the protective system will dicsonnect a short circuit within one second at the latest. In most cases only a fraction of a second is required.

Switching processes within the network may also introduce impulse-type noise. Three-pole disconnects operate comparatively slowly. If they are used to tie in a short line section, the mean noise pulse density approaches 600 pulses/s. Depending on the design of the disconnects, the disturbance may last up to 2 s. Circuit breakers, which reconnect also live line sections, perform switching functions within a far shorter time. The impulse density may here be still higher, but the disturbance will not last longer than some 25 ms.

On the line, the amplitudes of these noise pulses may reach the magnitude of the operating voltage. They are limited in the coupling circuit of the carrier terminal [17]. Special transmission methods are employed to isolate the carrier frequency receivers (see page 127) from the effects of the remaining high noise pulse voltages.

Systematic investigations have been conducted in many network protection systems to obtain information on the dependability of carrier signal transmission in the presence of power line failure, and data on the behavior of the protective systems have been collected over prolonged periods [10]. A CCITT recommendation, dated September 1958, specifies a disconnect time not exceeding 0.2 s for all hazardous voltages induced in the communication systems of the Post and Railways by faults in the 220 kV network with solidly grounded neutral [41]. The results of this measure are very promising so that this type of carrier relaying has found widespread use.

Carrier terminals should be designed not only for their most frequent application, i. e. for use on a line terminated by two stations, they should also be capable of serving lines to which three or even more stations are connected. In such cases the problem of network protection is normally solved by providing for each three-phase system one carrier transmitter at each end, and one carrier receiver at every other end of the line (Fig. 45).

In the case of a power line and two spur lines, one transmitter and three receivers are required at each end of the line. For two three-phase systems, a total of two transmitters and six receivers have to be provided. These requirements are reduced if protective relaying is simplified because one of the terminal stations need not transmit a signal. This may be the case where one terminal station is merely a load and no feed-in point. It may happen that a station is by-passed by one of the two three-

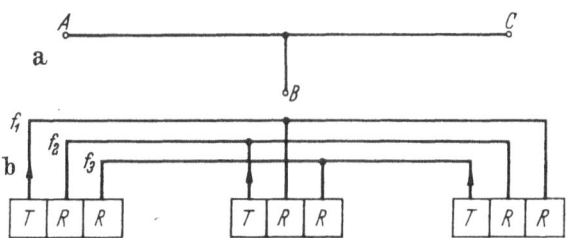

Fig. 45a and b. Protective relaying channels for a line section with three terminal points.
a) Location of stations A, B, C; b) Arrangement of carrier channels f_1 to f_3

phase systems and the second system is looped in or connected by a tie line, while the next station is by-passed by the previously looped in system and connected by a tie line or a loop to the other system (Fig.46). This example may serve to give the reader an idea of the number of

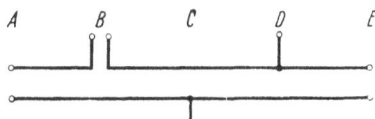

Fig. 46. Power line sections with more than two terminal points

possible combinations of transmitters and receivers. It is obvious that here the carrier terminals for protective relaying must be highly flexible and capable of easy adaptation to new conditions, as may be created by network modifications. This is to say that only single-purpose carrier terminals will suit the purpose.

If there are only two terminal points per line section, multi-purpose equipment for alternate transmission may also be employed to advantage. Multi-purpose terminals transmit the protective signals in a superimposed signaling channel, which must be adapted to its special function of protective relaying and which is therefore different from a telemetering or remote control channel. Two channels are superimposed on the speech channel if two three-phase systems have to be protected. Employing a suitable modulator circuit, that portion of the total transmitter power which is made available in a multi-purpose terminal for a superimposed protective signal channel, should suffice for dependable signal transmission over the usual line sections with two terminal points. Furthermore, the transmitter power can be raised for the signaling channel as the speech channel is interrupted for the brief duration of protective signal transmission so that the full key-down power is temporarily made available for protective relaying. Since this interruption lasts only for fractions of a second, it will hardly be noticed by the conversing parties.

Multi-purpose equipment for alternate operation transmits the protective signals invariably with the full transmitter power, which is here made available either for speech or for protective signal transmission. The changeover takes a little time, as does the transition to full key-down power in multi-purpose terminals. This results in a slightly increased tripping delay, compared with the permanently assigned protective signal channels provided by single-purpose equipment, or by multipurpose equipment without full key-down power for signaling.

For single-pole breaking in connection with reclosers in networks having a solidly grounded neutral it would be desirable to have a separate carrier channel for each phase. This amounts to 3 channels for each direction of transmission in a single three-phase system and to 6 channels for a double system. The resulting abnormally high equipment and frequency requirements can be curtailed by a phase selector built into the protective relays so that a single carrier channel will in most cases suffice for one three-phase system and one direction of transmission.

6.6 Portable Carrier Telephone Terminals for Line Crews

At a time when speech was the only means of intelligence transmitted over power lines to assist electric system operation, portable carrier terminals came into general use for establishing contact with line crews. These terminals were designed to provide communication with the nearest attended station in which a fixed-plant carrier terminal was installed. Originally, contact was to be established merely with the operators in the nearest station. At a later time, facilities were to be provided for through-dialing within the entire carrier communication network.

In densely populated industrialized countries other means of communication were available for line crews right from the start so that portable carrier terminals had to be used only in isolated cases. Today, distances of up to 25 miles are spanned by portable radio sets. The original problem of contacting a line crew over large distances, using a power line as the only communication path available, no longer exists except for some large and sparsely populated countries.

In some special cases portable equipment is used in temporarily attended stations to save the cost of fixed-plant carrier terminals. Everyone who wants to talk from such a station has to bring along his own carrier equipment and the associated power supply. In most cases this equipment is mounted in a motor vehicle and can be tuned to a number of different frequency pairs. The coupling circuits are here permanently installed and the terminal fittings on the low voltage side of the line tuning units are connected to a wall outlet. Coupling capacitors are occasionally designed to function as capacitive voltage dividers and the operating voltage for the carrier terminal can then be tapped from this source.

The most frequent application, however, is coupling a portable carrier terminal en route to a disconnected and grounded power line, to permit conversation between line men and a supervisor. As the power line is grounded, portable wave traps (Fig. 24) must be interposed in the grounding wires. Since communication with the line crew is to be maintained until the repair work has been completed, i. e. until after the ground wires have been removed and the line has been reconnected to power, some other means of coupling must be employed, because a portable coupling capacitor rated for the full operating voltage would be too heavy and unwieldy, not to speak of the shock hazard created by its exposed connecting leads. Moreover, supervisors are reluctant to have a line reconnected to power at a time previously agreed upon over the telephone as there is actually no assurance that the telephone system has been disconnected between the last conversation and the time fixed for energization. Inadequate as it may be, the only comparatively convenient solution left is the use of a coupling antenna. Its poor performance in carrier transmission is here not felt to be as great a handicap as in fixed-plant installations which are expected to afford a much greater transmission range, good transmission conditions for long-haul communication, and a higher dependability under abnormal weather conditions.

A portable carrier telephone terminal should be capable of service within a large area of the network, i. e. its transmitter and its receiver should be tunable to a number of frequency pairs used in different districts. Besides, the send and receive frequencies of tuned equipment must be interchangeable, to permit conversation in both directions within telephone districts operating with permanently assigned carrier frequencies, i. e. to permit conversation with single-purpose equipment designed for point-to-point or radial traffic. Portable carrier terminals cannot be used in telephone districts where the carrier frequency is continuously transmitted, point-to-point and waystation circuits are typical examples, because interference phenomena and disturbance by superimposed signaling channels are here unavoidable. Similar conditions obtain in carrier districts employing multi-purpose equipment for alternate transmission. In view of these limitations, portable carrier terminals, even the most simple versions operating with double-sideband transmission, are used only in very rare cases.

A way-out of these difficulties, permitting also the use of portable equipment working with any type of modulation, consists in abandoning the idea of employing only one equipment for interoperation with a fixed-plant carrier terminal. Things are made much easier by using a pair of carrier terminals permanently tuned to two service call frequencies. One carrier terminal is set up at the site of repair work, and the other carrier terminal at the desired distant station. The equipment at one end is coupled by way of a permanently installed coupling circuit, while the equipment at the other end is coupled by means of an antenna. The telephone district thereby provided for the line crew is completely independent of the mode of operation of the fixed-plant carrier terminals. One important point, however, must not be forgotten in this connection. The frequency allocation scheme must make provision for these service call frequencies. Under certain circumstances this might involve appreciable extra costs, since anti-resonant traps have then to be provided in the entire carrier communication network even where resonant traps would otherwise suffice for the operation of the fixed-plant carrier terminals.

7. Carrier Communication Terminals

Development in power line carrier communication has for many years been guided by the assumption that a single communication path between the terminal stations of a power line would be all that is required. Even where two or three telephone circuits were found to be desirable at a later time, recourse was again had to single-channel telephone terminals, which were then connected in parallel to the input of the coupling circuits. These carrier terminals were designed exclusively for telephone conversation. When remote supervision and control entered the picture at some later date, the first carrier terminals built were again of the single-purpose type and their transmission band was chosen so as to fit into the allocation scheme established for the speech frequencies.

Multi-purpose equipment for the simultaneous transmission of speech and supervisory control signals was subsequently developed with the objective of distributing the basic costs of the carrier channel to both functions, in other words, for reasons of economy. Attempts were also made to explain this trend in the development by technical reasons, asserting that the requirement for signaling channels was low and that frequency positions could be saved. The economical as well as the technical reasons can be accepted only with some reservations. Single-purpose and multi-purpose carrier terminals both have their place in carrier communication. The question whether the one or the other will offer more advantages is decided exclusively by the transmission problem which has to be solved in each specific case (see page 68). It might happen that both versions lend thenselves equally well to a certain application.

Multi-purpose equipment for alternate operation transmits speech as long as pulsed information must not be impressed upon the carrier instead of the speech currents. Pulse transmissions of some length pose no special problems to the carrier frequency section of the equipment. Such problems have to be dealt with by the automatic switching units connected. This type of multiple utilization is normally restricted to applications where the transmission of the pulsed information takes very little time, as in protective carrier relaying.

Attempts to develop a uniform version of a single-channel terminal for optional use on normal communication lines and on power lines proved to be unsuccessful because of the greatly differing transmission characteristics of these two communication paths. The specific require-ments of remote supervision and control were another point which had to be taken into account. The sum total of all these requirements led to the development of equipment in each of the aforementioned three categories which, with the exception of some common circuit components, differs greatly from the equipment versions designed for use on postal lines. A particularly typical case are the single-purpose multi-channel terminals for supervision and control.

A grouping of the carrier terminals as to their application and the number of channels transmitted results in a survey (Fig. 47), which today no longer includes single-channel single-purpose equipment for remote supervision and control purposes, because it is wasting bandwidth. On the other hand, it includes multi-channel telephone terminals, which were rarely used in the past. Attempts are again being made to apply the multi-channel carrier terminals designed for postal applications, but their adoption on a larger scale is again hindered by the previously mentioned difficulties, inherent in the transmission characteristics of power lines, and by the specific problems arising in power system operation.

In view of the crowded frequency band it would be desirable that all multi-channel carrier terminals operate on the single-sideband principle, while single-channel terminals, the almost exclusive choice for voice communication, should be operated on the double-sideband or single-sideband, or maybe on the frequency-modulation principle, depending on the transmission problem to be solved.

In some countries the prevailing tendency is to construct power line carrier networks only with such carrier channels as require no differentia-

tion as to their application in the terminal equipment. What is wanted
are 4-wire circuits between stations and weak channel groups between
nodal points. These circuits, similar to cable lines, should be optionally
available for telephony, supervision and control, or teleprinter traffic.
Such carrier networks permit only point-to-point traffic in the various
districts, because these demands cannot be met neither in frequency
interchange districts nor in waystation districts. The carrier terminals
contain only the carrier section with four-wire output, and are therefore
pure channel units. All components required for a specific application,
such as multi-purpose or alternate-use options, are here omitted and the
only distinction is between single-channel and multi-channel terminals.
All components required for using such a terminal for telephony or for
supervision and control are accommodated in accessory units, which are
derived from other applications.

Any special developments initiated by the use of the power line as a
transmission medium can be limited to the carrier section. This makes
for a reduction in the number of different versions. In toto, however, this
method will not result in any appreciable saving in development costs
because, due to the specific problems encountered in remote supervision
and control, the design of new special versions of extra equipment
becomes unavoidable.

Constructing a carrier communication network with the aid of
channelizing equipment and with accessory equipment adaptable to a
specific application (telephony, supervision, control) and to the mode of
operation (single-purpose, multi-purpose, alternate -use) means a higher
flexibility in the case of network expansion and network modification.
On the other hand, the higher installed costs per station, caused by the
distribution of functions to several constructional units, tend to set a
limit to their application. This applies in particular to smaller communication systems. Equipment combining the components required for the
various functions in a single constructional unit will therefore hold its
place. A recent trend in equipment design has led to a carrier terminal
which can be equipped for a specific application and a specific mode of
operation and be readily changed-over to another option if and when so
desired. This will offset the extra cost involved by the separation into
two constructional units, the carrier section and the accessory section.

The measures so far described for counteracting a high noise level on
power lines consist in stepping up the send power and selecting a suitable
transmission method. Compandering a circuit is another way to achieve
this purpose. The compandor is supplied as an accessory to the carrier
terminal. Its compressor is connected to the output of the voice frequency section ahead of the modulator to restrict the volume range
of speech signals, while the "expandor" in the receiving equipment is

	Single-channel operation		Multi-channel operation			
	for	Section No.	for	Section No.	for	Section No.
Single-purpose equipment	Telephony	7.1 a	Supervision and control, protective relaying, high-speed switching	7.1 b 7.1 c	Telephony	7.4
Multi-purpose equipment	—	—	Telephony and supervision and control	7.2	Telephony and supervision and control on one channel	7.4
Multi-purpose equipment for alternate transmission	Telephony or supervision and control	7.3	—	—	Telephony or supervision and control on one channel	7.4
Transmission method:	Double Side Band (DSB) Single Side Band (SSB) Frequency Modulation (FM)				Single Side Band (SSB)	

Fig. 47. Survey of power line carrier communication terminals

intended to restore it. A circuit can be compandered no matter whether the signals are transmitted with amplitude modulation or with frequency modulation. The objective is to raise low speech levels, which would make the transmission system most sensitive to noise disturbances, in the compressor of the transmitter so that these low speech signals are transmitted at an increased level over the line. The expandor in the receiving equipment serves to reduce the level raised in the compressor and, with it, the level of the noise signals picked up along the transmission route.

A compandor may be supplied as a separate accessory unit for improving the performance of existing carrier terminals in older communication systems, or it may be supplied as an integral part of the carrier terminal. In the latter case, its function has been considered in fixing the design objective, which then provides for a lower transmitter power (far below 10 W) to reduce the complexity of the send amplifier and its power supply, since these items account for a substantial portion of the total manufacturing costs.

A compandor has, unfortunately, the disadvantage of introducing non-linear distortion in a transmission section, since its operating principle is based on the use of circuit elements having non-linear current-voltage characteristics. Although the amount of distortion generated will hardly be annoying in a single section or in two tandem-connected sections, a circuit consisting of a larger number of cascaded links might be degraded so that the non-compandered circuit would give better results. It is therefore advisable to use compandors only at the beginning and at the end of such a long-haul circuit, and to disconnect them in the intermediate stations. A compandor cannot be regarded as a generally applicable means of ensuring satisfactory transmission quality as is a sufficiently high transmitter power, or an appropriate transmission method. Its use is indicated merely for improving medium-grade conversation. It is not capable of improving conditions in marginal cases, i. e. in cases where the signal-to-noise ratio is abnormally poor.

7.1 Single-Purpose Equipment

The description of this type of equipment will be limited to carrier terminals serving three different transmission tasks (section 7.1 a–c, Fig. 47). These are telephony and two sectors in the field of remote supervision and control. Older equipment versions will not be discussed, although they have not yet been entirely withdrawn from service. The three transmission methods (DSB, SSB and FM) could be employed, in principle, both for telephony and for supervision and control. Nevertheless, it is only in telephony that all three methods are extensively used.

a) Telephone Terminals

Until about 1940, amplitude-modulated DSB carrier terminals were almost exclusively used for speech transmission. Since the double-sideband technique affords a less complicated circuit design, the DSB terminals remain the preferred choice unless spectrum crowding or a high noise level tip the scales in favor of the single-sideband technique, i. e. chiefly in networks up to 110 kV.

The basic circuitry of DSB carrier telephone terminals (Fig. 48) is the same for the large majority of the different makes. However, as far as

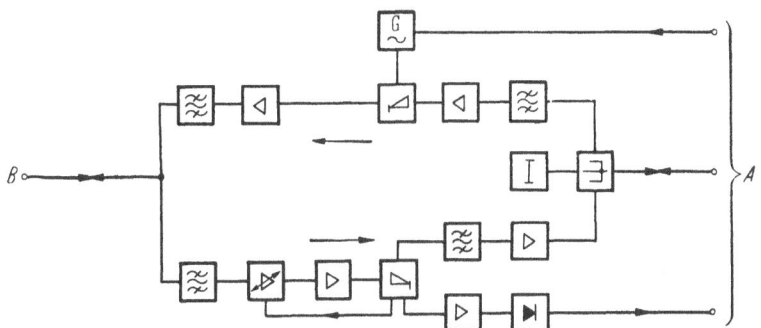

Fig. 48. Basic circuit arrangement of a DSB carrier telephone terminal.
A Station side (via relay unit); *B* Line side. For explanation of symbols refer to Fig. 53

the circuitry of the various constructional units of a carrier terminal is concerned, manufacturers are following widely differing design principles.

The VF speech currents to the HF send branch and from the HF receive branch of the carrier terminal are transmitted over a two-wire line between the telephone station and the hybrid. The outgoing dial pulses (sent in the form of a break in the DC loop current by the dial switch contact) as well as the incoming ringing signals (AC current pulses from a ringing current source inside the carrier terminal) are normally sent over the same wire pair which is used for speech transmission. A telephone station is here connected to the carrier terminal in the same way as to an automatic exchange.

The "hybrid" provides for the transition between the two-wire circuit, connecting the telephone station to the carrier terminal, and the long-distance path over which the signals are transmitted on a four-wire basis (Fig. 49). The speech currents arriving from the telephone station are applied to the transmitter by way of a differential transformer, while the currents supplied by the receiver are passed on to the station loop. To prevent the incoming speech currents from flowing back through the own transmitter, the impedance of the balancing network must be equal to the impedance of the station loop. The incoming speech currents are thereby distributed equally to the telephone station and to the balancing network. The currents arriving from the telephone station are also halved, so that one part of the energy flows into the balancing network. This hybrid is omitted in carrier terminals having a four-wire voice frequency output. If a carrier terminal designed for a two-wire output has to be connected to an exchange system operating on a four-wire basis, the hybrid is disconnected.

The VF amplifier raises the level of the outgoing speech currents so that the desired depth of modulation (normally not more than 80%) can be obtained. Furthermore, it contains an amplitude limiter intended to prevent overloading of the power amplifier so as to assure its optimum

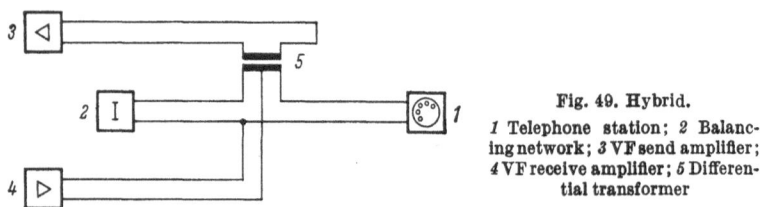

Fig. 49. Hybrid.
1 Telephone station; *2* Balancing network; *3* VF send amplifier; *4* VF receive amplifier; *5* Differential transformer

utilization. In the modulator, the carrier current generated in the HF oscillator is modulated with the speech signal. The send amplifier raises the level of the modulated carrier. A certain portion of the send amplifier output is dissipated in the send filter. The HF send filter and the HF receive filter fix the two frequency positions which have been assigned to the channels of the carrier terminal.

The carrier frequency transmitted by the distant terminal in the opposite direction is applied to the demodulator of the receiving equipment by way of the HF receive filter, the automatic volume control, and the HF amplifier. The automatic volume control circuit ensures a HF voltage of constant amplitude at the input terminals of the demodulator. Behind the demodulator the incoming message is branched off into a ringing path (amplifier, detector, call receive relay) and a speech path (voice bandpass filter, amplifier, hybrid).

The mode of operation of a carrier telephone terminal varies with the type of telephone district. In point-to-point traffic, for instance, the carrier frequency is in most cases transmitted continuously in both directions to enable the volume control circuits at both ends to adjust the sensitivity of their detector circuits continuously to the attenuation conditions on the line. The same applies to way-station traffic. With frequency interchange traffic on the other hand, the receivers are continuously switched on in the standby condition so that a call can be received at any time. During this standby condition the receivers are always conditioned for maximum sensitivity. The transmitter, however, is connected to line only when a call is to be transmitted. No carrier current is transmitted during the rest condition. When the carrier frequency arrives at the beginning of a call, the automatic volume control will adjust the receivers for a lower sensitivity, depending on the distance between the conversing telephone stations and on the prevailing weather conditions.

Carrier telephone terminals have an average transmission range equalling 50 db to 60 db. In practice, however, this range can hardly be obtained in view of the noise level (Appendix 9.7). The control range amounts normally to some 35 db at an accuracy of up to ± 0.8 db within this range.

By the time dial pulse transmission begins, the receivers have adjusted themselves to the prevailing level and a possible distortion of the dial pulses is thereby avoided. The transmitter of the called carrier terminal is switched on as soon as a carrier signal has been received. In frequency interchange districts only the transmitter of that carrier terminal is switched on, which has actually been contacted by means of the call number. The HF carrier subsequently transmitted is modulated with a buzzer tone, to provide the calling station with an indication whether or not the called terminal is free. In other words, the calling party receives a ringing tone. At the end of the conversation, the connection is taken down by restoring the handset. Whether the called or the calling party is the first to cradle the handset, or whether one of the two parties fails to replace the handset is of no consequence for the release process. If the carrier terminal is connected to an automatic exchange, the latter must provide for the switching criteria required for releasing a connection, no matter in which direction the connection had been established.

In telephone districts where the carrier frequencies are continuously transmitted, the receivers are already matched to the existing level before the dial process begins. In other words, these receivers are less sensitive to HF noise generators than the receivers in frequency interchange districts. Various precautions can be taken to prevent HF noise from causing false seizure. While these measures can be adopted for any receiver, the extra cost mostly involved will limit their application to the receivers of frequency interchange terminals, which are not continuously adjusted to the existing line attenuation.

Number plate dialing is the rule also in telephone districts including only two telephone terminals, with one telephone station connected at either end. In such cases the transmission of the carrier alone would suffice for calling the distant station. A call might be simulated, however, by a HF noise pulse of sufficient amplitude. A certain safeguard against false seizure can be provided if the carrier terminal is designed so that it will respond only to a series of dial pulses.

Another way of preventing a receiver, whose amplification has been pushed up to the highest degree, from responding to brief HF interference pulses consists in providing a delay network, which will interpret a HF current as a call signal only if this current lasts for more than about 300 ms.

A still more effective method of protecting a ringing signal receiver against false seizure consists in radically reducing the bandwidth of the carrier channel during dialing, as compared to the bandwidth made available for speech transmission. An adapter filter with a narrow pass-band range is then connected ahead of the dial pulse receiver. This filter is automatically cut out as soon as the receiver has adjusted itself to the prevailing level and the dialing process has been completed. The sensitivity of the HF receiver is thus reduced and the 5 kHz voice band can be accepted.

For operation in the frequency range assigned to power line carrier terminals these adapter filters must be crystal-stabilized filters. This involves the necessity of using also crystal-controlled oscillators for the HF transmitters, since the temperature response of the tuning media normally employed for transmitters might produce an undue departure of the carrier frequency. Assuming a 5 kHz transmission band and a bandwidth of 150 Hz for the adapter filter, which affords an attenuation of some 26 db within the frequency range to be blocked, a decrease of the noise voltage in the order of 12 db can be achieved in the signaling channel at a nominal transmission frequency of 150 kHz.

Adapter filters for dial pulse receivers are not necessarily a standard item of equipment of carrier telephone terminals. However, in all cases where HF interference pulses occur quite frequently because high loads must often be switched at high voltages or because a particular region is frequently hit by thunderstorms, the use of such filters is definitely indicated to minimize the hazard of false seizure. The receiver will then respond only to that portion of the noise volume which remains within the 150 Hz band. This method affords a high measure of protection against interference pulses.

Another means of preventing false seizure is "voice frequency dialing". The dial pulses are here no longer transmitted by on/off keying of the carrier, but by modulating the carrier with a frequency falling within the range of the voice band. This can be done because the dialing process invariably precedes message transmission and because the dialing frequency remains disconnected during the conversation. The necessary filters need not be crystal filters. The filters usually employed in VF telegraphy will do.

Voice frequency dialing is something particular only for those DSB carrier terminals which transmit the dial information by on/off keying of the carrier. It is a standard feature in multi-purpose terminals which, in addition to speech, continuously transmit supervision and control channels. The same is true of single-purpose equipment transmitting a pilot frequency for continuous volume control. A frequency outside the voice band ist mostly used in these cases (out-of-band dialing), but this is of no consequence as far as protection against false seizure is concerned.

The forementioned amplitude-modulated DSB carrier telephone terminal was the first type of carrier terminal developed for use on power lines. Since its introduction in 1920 it has seen many improvements and is still used. From about 1940 onwards the ever increasing frequency shortage compelled the system operators to adopt SSB carrier terminals for the transmission of speech over power lines. Although SSB terminals had been operated for some years on postal lines, the power companies were reluctant to adopt such terminals for economical reasons. Whereas in postal networks the higher installed costs of SSB terminals can be apportioned among a large number of channel groups, the full impact of this change-over was clearly felt by the power companies whose requirement, at least at that time, was almost exclusively for single-channel carrier telephone terminals. This is all the more annoying if a rigid separation between send and receive channels must be abandoned in favor of a variable separation, to retain the flexibility in frequency assignment which is highly desirable in intermeshed networks.

Power line carrier telephone terminals for single-sideband operation within a frequency region extending up to 500 kHz must necessarily work with two-stage modulation. As distinct from operation in the lower frequency ranges, it is here no longer possible to use the normal circuit elements for suppressing one sideband and the carrier immediately in the HF position. The reasons are explained in the following:

Assuming a lower voice band limit of 300 Hz the spacing between the two lower frequencies of the sidebands amounts to 600 Hz (Fig. 50). Expressed in percents, this spacing equals 5 % at a low carrier frequency of, say, 12 kHz. For operation at 120 kHz, this figure drops to 0.5 %. If a single sideband is to be transmitted, both the carrier and the other sideband must be eliminated. This can be done with normal bandpass filters if the spacing between the two sidebands amounts to 5 % of the carrier frequency. To cope with a spacing of only 0.5 %, the components of the bandpass filters would have to satisfy higher demands as regards stability and low-loss characteristics, so that crystal filters would have to be used.

At frequencies ranging between 15 kHz and 500 kHz, the sideband separation will lie between 4 % and 0.12 %. To avoid the necessity of using crystal filters, recourse is had to two-stage modulation. In the first

modulation stage an intermediate frequency of, say, 12 kHz is modulated
with the voice band. The spacing between the two sidebands amounts
to 5% of the intermediate frequency so that elimination of the intermediate frequency and of one sideband can be accomplished with simple
means. The carrier frequency, which is assumed to be 120 kHz for the
purpose of this example, is modulated only with one of the two sidebands,
say with the upper sideband. This sideband appears in the normal position between 12.3 kHz and 14.4 kHz. In the HF position, the lower

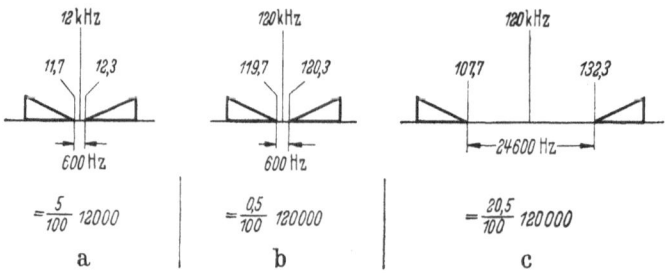

Fig. 50a–c. Relative spacing between sidebands and carrier at different carrier frequencies.
a) Transmission at low carrier frequencies; b) Transmission at high carrier frequencies, single-stage
modulation; c) Transmission at high carrier frequencies, two-stage modulation

sideband is obtained in its inverted position between 105.6 kHz and
107.7 kHz, while the upper sideband appears in its normal position
between 132.3 kHz and 134.4 kHz (Fig. 50c).

The spacing between the two sidebands amounts to 24.6 kHz, that is,
to 20.5% of the carrier frequency, so that elimination of one sideband
and of the carrier can be achieved with simple means also in the HF
position.

The crystals required for stabilizing the intermediate frequency and
HF oscillators can be largely standardized for the intermediate frequency
range of transmission systems. It will therefore suffice to produce crystals
for a small number of intermediate frequencies. On the other hand,
crystals of many different frequencies are required in the HF band to
shift the transmission band from its intermediate frequency position
into the desired HF position within the frequency allocation scheme. An
overall range extending from 30 kHz to 490 kHz, for instance, affords
184 different channel frequencies in the case of a 2.5 kHz allocation
scheme. Thus, the HF crystal must be produced for one of the 184 possible
frequencies provided by the allocation scheme. To reduce the number of
crystal frequencies in an effort to simplify manufacture and spare parts
keeping, crystals in the megacycle range are also used. Their frequency is
stepped down at various ratios by frequency dividers, e. g. at the ratio
2 : 1 up to 16 : 1. A common crystal frequency will then do for a number

of channel frequencies. As an incidental advantage the stability of crystals increases with the frequency so that carrier frequency oscillators may then be operated without the need for thermostat-controlled crystal ovens.

Where two contiguous frequency bands are to be used for the two directions of transmission, a common intermediate frequency oscillator may be used for the converter stages of both directions, as well as a common HF oscillator (Fig. 51, top).

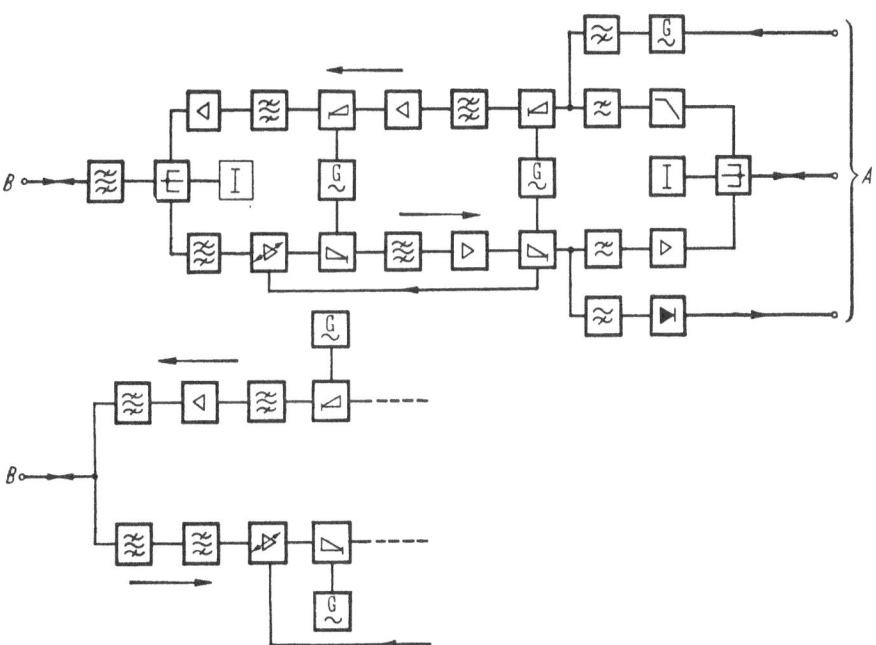

Fig. 51. Basic circuitry of an SSB carrier telephone terminal.
A Telephone station side; *B* Line side. (For explanation of symbols refer to Fig. 53)

Since neither the carrier nor the intermediate frequency is transmitted, the dial signals must be applied in voice frequency form. It would here be possible to adopt in-band dialing because dialing and speech transmission are consecutive processes. However, a voice frequency outside the speech band is already continuously transmitted as a pilot frequency for the automatic volume control. This frequency is transmitted during conversation and during silent periods. The dial signals can therefore be transmitted either over a second frequency outside the speech channel, or by keying the pilot channel. In the latter case, the time constant of the automatic volume control must be chosen so that the inter-digit intervals will be bridged.

The speech currents arriving from the telephone station reach the intermediate frequency modulator by way of a hybrid, an amplitude limiter and a highpass filter. An intermediate frequency filter allows only one sideband to pass and to proceed, by way of an intermediate frequency amplifier, to the HF modulator. In the HF position it is again only one sideband which is applied to the output amplifier by way of an HF filter.

In the incoming direction, the sideband transmitted from the distant station is applied to the HF demodulator by way of an automatic volume control and an amplitude limiter. As the incoming signal is translated from the HF position to the intermediate frequency position, the residual carrier and one sideband are absorbed by an intermediate frequency filter, so that only the second sideband will pass on to the intermediate frequency amplifier and the following second-stage demodulator, where the intermediate frequency is added to convert the frequency band down to the voice frequency position. The speech currents are then fed on to the telephone station through a VF bandpass filter, a VF receive amplifier and the hybrid.

In an SSB carrier terminal designed for contiguous band working, the two frequency bands of the two directions of transmission are separated by an HF hybrid, which is connected to the HF input of the carrier terminal. If an SSB telephone terminal is to be designed so that the gap between the two frequency bands assigned to the two directions of transmission is not permanently zero, but variable as desired, the HF hybrid must be replaced by two HF frequency separating filters. Besides, two HF oscillators tuned to different frequencies will then be required (Fig. 51, bottom).

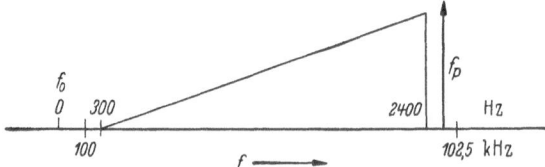

Fig. 52. Position of zero frequency for an SSB telephone channel (300 Hz to 2400 Hz) within a 2.5 kHz allocation scheme.

f_0 Zero frequency; f_p Pilot frequency

The pilot frequency may be transmitted in the form of a partly suppressed intermediate frequency, that is, a carrier of reduced power may be transmitted together with the speech signals. This will save the cost of a separate pilot oscillator and filter, but monitoring of such a single-sideband transmission will then be just as easy as monitoring of a double-sideband transmission. To avoid this, a separate pilot frequency oscillator may be used, which provides a frequency above the upper speech frequency. If this is done in a 2.5 kHz allocation scheme, it will be necessary to shift the "zero frequency" to the adjoining frequency position (Fig. 52), because a sufficient separation must be provided between the upper cut-off frequency of the speech band, the pilot frequency, and the cut-off frequency of the respective position, to permit the use of uncomplicated filter circuits. The extra cost of a "residual carrier trap" can be avoided by arranging the zero frequency inside the frequency position and by clipping the speech band at about 2100 Hz. Bearing in mind the CCITT recommendations regarding the quality of speech transmission, this solution should be avoided whenever possible (see page 66).

A certain amount of protection against false seizure of the ringing signal receiver is inherent in the operating principle. It is afforded partly by the limitation of the noise volume in the ringing signal channel, which is by far narrower than the voice channel, and partly by the continuously effective volume control, which keeps the sensitivity of the receiver always at the actually required level.

Frequency modulated carrier terminals intended for operation over power lines must be designed so that their frequency bands coincide with the frequency allocation scheme, in other words, their frequencies should fit into the positions of a DSB allocation scheme. The frequency swing is therefore approximately ± 2 kHz (Appendix 9.6). Even within so narrow a frequency band, frequency modulation offers a signal-to-noise advantage of about 9.5 db over a DSB carrier terminal of equal carrier power, employing a depth of modulation of 80%. If these two types of equipment are engineered for the same signal-to-noise ratio, that is, for the same transmission range, the frequency-modulated equipment can operate at a level which is lower by 9.5 db and can therefore be equipped with a smaller send amplifier.

The basic circuit layout (Fig. 53) resembles that of a DSB carrier terminal. The speech signals are applied to a hybrid with balancing network by way of a two-wire line. The VF amplifier raises the amplitude of the speech signals to the level required for modulation. A series-resonant circuit accentuates the higher frequencies (pre-emphasis) so that they are transmitted at the same frequency swing as the medium frequencies. The VF signals applied to the modulator control the carrier frequency supplied by the HF oscillator with a maximum frequency swing of ± 2 kHz. The amplitude of the HF signal is subsequently limited and the signal is brought to its output level in an HF amplifier. The signal is then fed on to the coupling circuit by way of the send filter and a matching transformer. The dial pulses are transmitted in a channel below the speech band. The ringing and busy signals are returned through a VF source.

The carrier frequency voltages arriving over the power line are applied, by way of the coupling circuit and the matching transformer, to the receive filters which reject the HF oscillations of the local transmitter and those of external transmitters. Thus, only the wanted frequency band is passed and subsequently amplified and limited so that the discriminator will receive a constant HF voltage also in those cases where

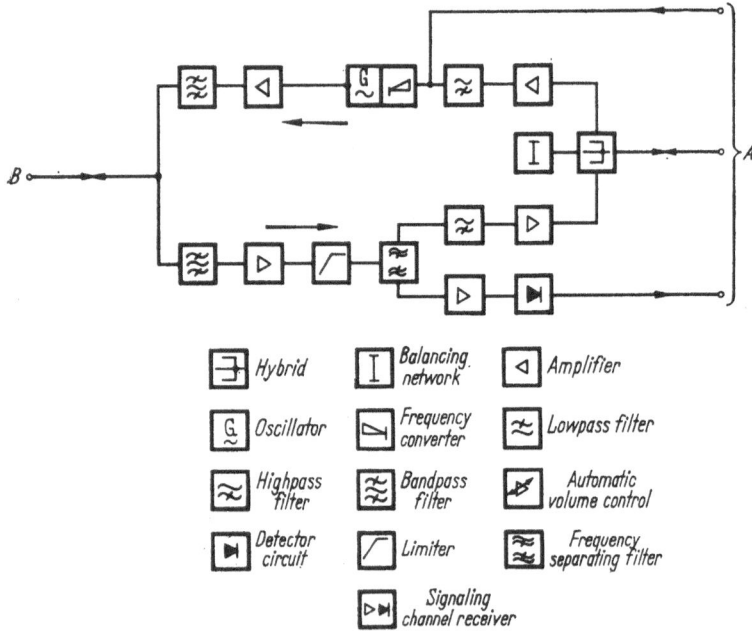

Fig. 53. Basic circuitry of a frequency-modulated carrier telephone terminal.
A Telephone station side; *B* Line side

the input level varies within rather wide limits. The discriminator converts the high-frequency oscillations of the carrier into the corresponding amplitude variations in the voice frequency range. The amplitudes of the higher transmission frequencies are reduced at the output of the discriminator to compensate for the accentuation effected at the send side (de-emphasis). The recovered VF voltages are raised to the desired output level in the following VF amplifier. The voice frequencies are then applied to the output circuit of the carrier terminal by way of a lowpass filter which rejects all frequencies beyond the 2400 Hz cut-off point.

The incoming dial pulses are branched off at the output of the VF amplifier and, after detection, applied to the dial pulse receive relay.

b) Carrier Terminals for Telemetering
and Remote Control

Carrier frequency terminals for supervision and control application are invariably designed as multi-channel single-purpose equipment, because the frequency space available within the allocation scheme is fixed by the bandwidth of the voice bands so that an entire group of narrow signaling channels can be accommodated. Some years ago, a carrier frequency, modulated by a telemetering signal, has been occasionally transmitted within a full frequency position. Today this waste of bandwidth can no longer be tolerated except in some smaller networks which still operate with earlier-type equipment. The same applies to carrier modulation with several voice frequencies and transmission on a double-sideband basis. DSB terminals for supervision and control used to be designed for the transmission of from 1 to 6 channels at a 120 Hz channel spacing. A larger number of channels could not be transmitted because the send power available for each individual channel was then too small to achieve an adequate transmission range. Although a larger number of channels may be obtained by using a high-power send amplifier an unnecessary waste of frequency space would result since, as in all DSB terminals, each channel is transmitted twice.

Carrier terminals for telemetering and remote control, which are in most cases required to accommodate many channels, are preferably designed for single-sideband operation (see page 109).

A description of the basic circuitry of carrier terminals intended for supervision and control applications might start with equipment types which could be classified as a modified version of a SSB telephone terminal whose carrier is modulated by VF signaling frequencies instead of the speech frequencies. It has been designed for waystation traffic, so that smaller groups of signaling channels can be added or dropped en route. What is important for an equipment operating with, for instance, 50-baud channels in a 2.5 kHz frequency allocation scheme is the fact that the sum of all channels must never exceed the figure 18 on any one transmission line section. A distinction is made between send terminals, receive terminals, combined send/receive terminals, and repeaters with channel adding and dropping facilities. These repeaters are designed for the conversion of the incoming carrier frequency so that the outgoing frequency band can be assigned a different position within the frequency allocation scheme. This is necessary to achieve sufficient decoupling between the input and the output of the repeater, so that its gain characteristic can be fully utilized despite the occasionally inadequate feedback attenuation of a power network station. A higher flexibility in frequency assignment is an incidental advantage of this measure.

In the send terminal (Fig. 54) the channel frequencies are generated by NF-oscillators. In the modulators, the amplitude of the individual frequencies is modulated by the external input devices in the cadence of the pulse trains. The aggregate signal, composed of the individual modulated channel frequencies, is shifted into the intermediate frequency

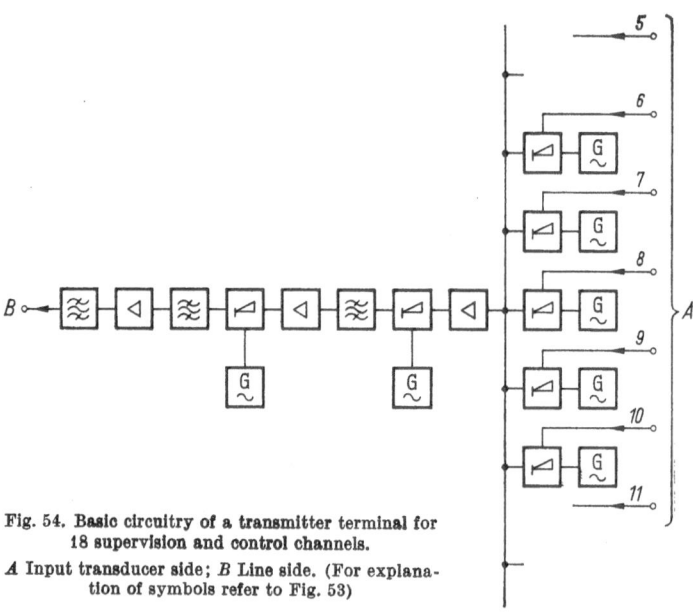

Fig. 54. Basic circuitry of a transmitter terminal for 18 supervision and control channels.

A Input transducer side; *B* Line side. (For explanation of symbols refer to Fig. 53)

range, brought into the HF range by means of filter arrays and a converter, and finally applied to the send amplifier. It is subsequently sent to the power line by way of the send filter and the coupling circuit. The send filter provides a high attenuation to all frequencies falling outside the transmission band.

One of the channels is also used for volume control at the receiving end. In the absence of signals, or in the event of a transgression of a predetermined no-signal period of, say, 500 ms, a steady tone signal is sent over this pilot channel. This is to ensure continued functioning of the volume control at the receiving end. If the pilot channel oscillator fails, an alarm will be initiated in the sending terminal.

In the receive terminal (Fig. 55), the composite signal is applied to one or to a number of channel group receivers by way of an attenuator with plugging options, designed to match the sensitivity of the equipment to the prevailing line conditions. Each channel group receiver is provided with a volume control to compensate level variations within the frequency group of the associated sending station. After two-stage demodulation, the frequencies are segregated by group filters and applied to the VF group amplifiers. Channel filters select the individual channel frequencies which, after detection, control the connected output units by way of transistor circuits. One of the channel receivers in each group is designed as a pilot tone receiver and supplies, in addition to the signal output, a control voltage for the volume control of the relative channel group. A monitoring circuit associated with each individual channel gives an alarm if the channel is faulted or no signals are received.

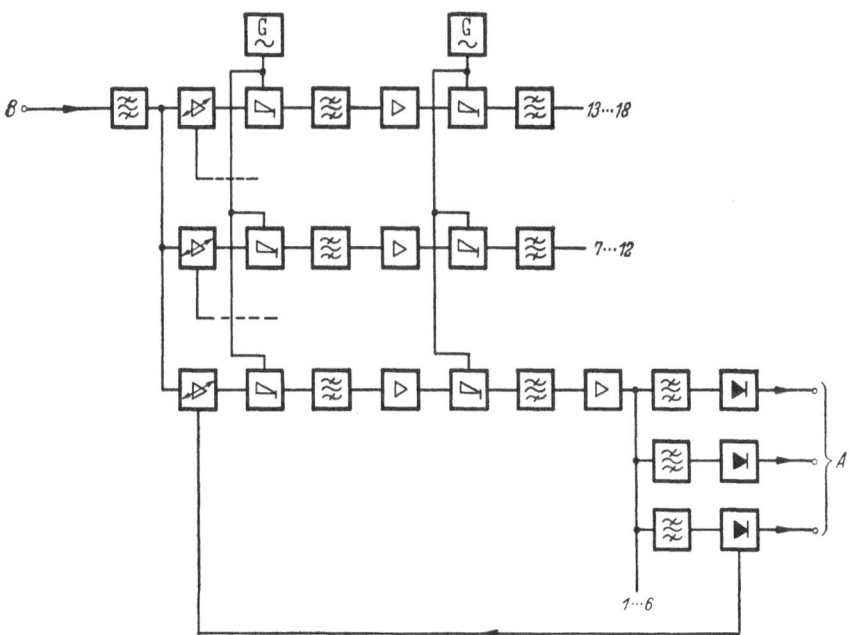

Fig. 55. Basic circuitry of a receive terminal for 18 supervision and control channels (in 3 groups). *A* Output transducer side; *B* Line side. (For explanation of symbols refer to Fig. 53)

The repeater cabinet with frequency converter stage (Fig. 56) consists of a receiving branch and a transmitting branch. Through-channels are transferred from the receiving to the transmitting branch in the VF range by means of bandpass filters. Channels to be dropped in this repeater station are branched off ahead of the converter stage. New channels may be added in the VF position.

If such a multi-channel carrier transmission system is rated, in compliance with regulations established in many countries, for a total transmitter output power of 10 W, the total peak power of 40 W can no longer be fully utilized if fewer than 4 channels are transmitted (Appendix 9.7).

c) Carrier Terminals for Network Protection and Direct Breaker Tripping

In network protection systems the emphasis is not only on a high measure of dependability, but also on particularly short delay times. This can be achieved only with channels having a corresponding bandwidth. Signal transmission in each individual line section proceeds normally

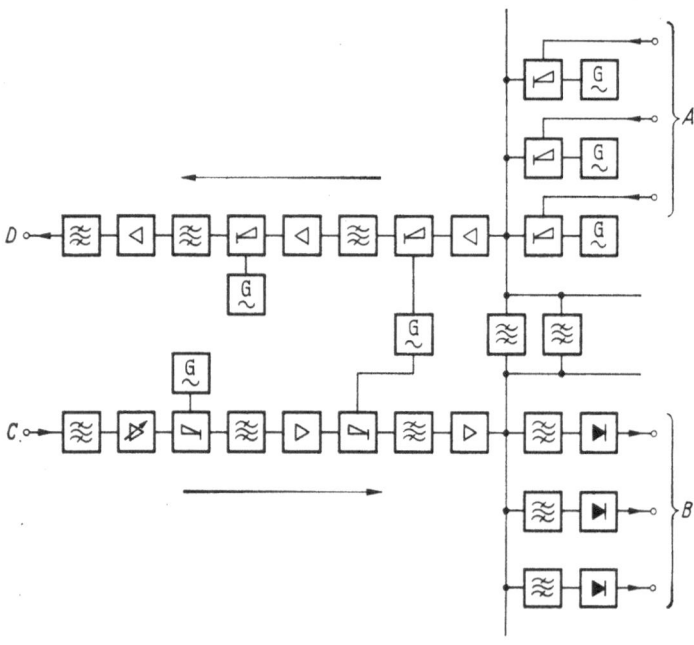

Fig. 56. Basic circuitry of a repeater with frequency converter stage. The functional units of the transmitting branch are the same as those employed in the send terminal (Fig. 54). The functional units of the receiving branch correspond to those of the receive terminal (Fig. 55).

A Input transducer side; *B* Output transducer side; *C* Power line side (incoming); *D* Power line side (outgoing). (For explanation of symbols see Fig. 53)

in both directions. Combined send/recieve units must therefore be provided as terminal equipment. The transmitted signal is intended to expedite (or to block) operation of the protective relays. In other words, a locally generated criterion must be available at the receiving station before the circuit breakers are tripped.

A similar problem exists if power breakers are to be tripped rapidly and securely by means of an uncoded signal. The most frequent applications are transformer stations or power stations being connected via spur lines to a power network which is only a few miles away from the stations. In this case the HT switch at the output of the station is to be saved and its function transferred to the next switch which is located at the other end of the spur line. When the protective relays of the generator or transformer operate, this remote switch should be tripped as rapidly and securely as if it were located in the station proper.

For these "high-speed tripping" applications signals are transmitted in one direction only. Thus, a transmitter is installed at the one and a receiver at the other end of the line. At the receiving end there is no locally generated criterion which would have any additional controlling effect on the remotely transmitted signal as regards tripping of the power breaker. The reliability of the protection system depends exclusively on the transmission equipment.

It seems reasonable to use a transmission equipment for network protection and for direct breaker tripping which can be employed as an adjunct to the three possible transmission media which are cable, carrier and radio circuits. The equipment intended to transmit network protection or direct breaker tripping signals should operate in the voice frequency range and be operated either directly on communication cables or, as a modulation adjunct to carrier or radio equipments. Cable connections are used for short distances, i. e. for network protection in medium voltage networks and city networks or for direct breaker tripping in high voltage networks. Carrier frequency and radio circuits, on the other hand, are employed in the case of great distances, i. e. chiefly for the transmission ot network protection signals.

The time available for the transmission of these signals is not more than about 10 ms. Adding the response time of the protection relay sets and the tripping time of the power breakers, the total time must not exceed fractions of a second if disconnection is to take place in time so that damage to the power plant can be avoided.

An equipment for transmitting network protection and direct breaker tripping signals must be immune to interference voltages, in particular to impulse-type voltages whose amplitude is great compared with the amplitude of the useful signal. To achieve this purpose, the following methods are employed in one version [14] of a VF equipment (Fig. 57):

a) To eliminate the influence of noise voltages on signal transmission, this equipment employs four-frequency shift keying (F 6 modulation), the signals being transmitted without breaks and with constant amplitude. One of the four possible frequencies is the guard tone. A protection signal is transmitted by shifting abruptly to one of the three working frequencies. The frequency thereby selected determines which of the two three-phase systems of a double line is to be disconnected or whether both systems are to be disconnected in common. The receiver detects only the frequencies and does not respond to changes in the signal amplitude. As a result of this, the amplitude of strong noise pulses is greatly limited at the full useful bandwidth so that these pulses are rendered ineffective.

b) Since only a single frequency is transmitted, the send power of the output amplifiers of interposed carrier or radio equipment can be utilized up to the maximum value.

c) As a result of amplitude limitation, level changes due to sudden variations of the attenuation of the transmission path are within wide limits without influence on signal transmission.

Reliability may be further increased by the following testing and supervising facilities:

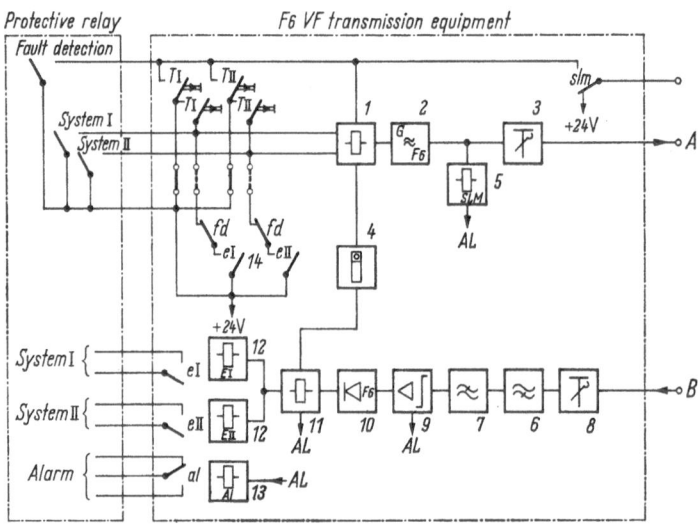

Fig. 57. Basic circuitry of a VF transmission equipment for transmitting network protection and direct breaker tripping signals with F 6 modulation (Siemens A.G.).

F 6 transmitter: *1* Frequency shift and adjustment; *2* Oscillator; *3* Attenuator; *4* Signal counter (2 each at send and receive sides); *5* Send level monitor; F 6 receiver: *6* Highpass filter; *7* Lowpass filter; *8* Attenuator; *9* Amplifier/limiter with level monitor; *10* Frequency separator; *11* Twinplex receiver (signal evaluation); *12* Relay output; *13* Alarm relay; *14* Contacts for "echo circuit", *A* To outgoing line circuit; *B* From incoming line circuit

d) The continuous transmission of one of four frequencies is utilized for the continuous supervision of all functional units of the equipment and of the transmission path.

e) Test buttons are provided to permit local testing of those components which are normally only operated by a tripping signal, i. e. testing without tripping of the power breakers.

f) In selective line protection a loop test can be performed in addition. A tripping signal is sent from the testing station and returned by the distant station. The tripping command is not effective unless a fault detector has actually operated at the distant station. Stepping of the

counters in the equipment at the testing station indicates proper operation of the double connection.

g) Level monitors tripping an alarm in the event of a decrease in or
absence of the level are included both in the send leg and in the receive
leg.

With four-frequency shift keying the transmission range or immunity
to noise pulses is better by 12 db than with two-frequency shift keying
if two protective channels are in both cases to be transmitted in a band
of about 2 kHz width. As a result, the send power with two-frequency
shift keying would have to be increased more than 16 times if the same
reliability as with four-frequency shift keying is to be achieved.

To permit independent transmission of the two switching signals,
two make contacts are required to control the transmitter. To transmit a
protective signal, the associated make contact operates a relay in transmitter (1) which shifts the F 6 generator (2) from the guard frequency to
the working frequency corresponding to the command. This is done by
changing the tank circuit inductance by means of the relay contacts.

The send level is adapted to the existing conditions by means of a
variable attenuator connected behind the generator. A send level
monitor (5) may be connected to the output of the generator.

Arranged in series at the input of the receiver are a variable attenuator (8), a highpass filter (6) and a lowpass filter (7) for frequency band
limitation. The variable attenuator (8) is used to adjust the sensitivity
of the receiver. In the following amplifier (9) the receive level is raised to
the required value and the amplitude limited. The level at the output of
the amplifier is independent of attenuation variations on the line. The
frequency separator (10) with its four bandpass filters is connected to the
output of the amplifier. The frequencies falling into the receive band are
rectified at the output of each bandpass filter and result in DC signals
of various magnitudes.

Associated with each rectifier output is a switching transistor with
relay in the signal evaluator (11). The four transistors and the frequency-
responsive elements are interconnected in such a way that only one
transistor can conduct at a time so that only one relay is operated. If
the useful carrier fails due to a fault in the transmitter or in the receiver
or as a result of a broken conductor, or if more than one discrete signal
is applied to the input of the four bandpass filters as a result of interference, the signal relays at the output of the receiver will be blocked
(security afforded by the selection of one out of four frequencies). One
of these relays is used for supervision while the other three relays pass
the protective signals on to the protection relays. For the direct tripping
of the power breaker a relay equipped with power contacts (12) is
required in addition.

Signal counters (4) on the receive and transmit sides record the incoming and outgoing signals and thereby permit a subsequent check of the number of tripping commands issued.

In the case of line sections fed from one end only, fault detection occurs only at one end. Tripping of the breakers is then initiated always by the station detecting the fault. The other station returns the protection signal to the fault detecting station ("echo circuit") so that the circuit breaker is also here tripped with a minimum of delay. However, twice the normal signal transmission time is here required.

A network protection or direct breaker tripping signal lasts only a few milliseconds. Moreover, cases where it must be transmitted are very infrequent, averaged over a year. The transmission channel is naturally used for telephony or for telemetering purposes during the rest of the time. For the sake of simplicity, however, the network protection circuit is permanently switched through via single-purpose terminals as long as a sufficient number of channels is available, i. e. in the case of short cable or radio connections. This is also usual practice in carrier communication over power lines. In the latter case, however, the increasing frequency shortage will frequently compel the system operators to ensure a more economical utilization of the available frequency spectrum, i. e. to adopt multi-purpose equipment for the transmission of speech, for instance, and for short-time changeover to protection signal transmission, if and when the need arises. This necessitates some accessories in the carrier equipment and in the F 6 modulator. The transmission delay of the signal is increased from about 10 ms to some 20 ms and the transmission range of the carrier terminals is reduced by about 12 db.

For direct breaker tripping, however, it should be made a point not to employ multi-purpose equipment so as to retain the advantage of increased transmission reliability.

7.2 Multi-Purpose Carrier Terminals for Simultaneous Transmission

While single-purpose carrier telephone terminals may operate on any one of the three conventional transmission methods, i. e. SSB, DSB and FM, there are two different reasons why single-sideband operation is preferred in multi-purpose terminals. One of these reasons may be the fact that a frequency allocation scheme has already been established at an earlier date for a large number of single-purpose carrier terminals operated on a double-sideband or on a frequency modulation basis. The broader frequency band of a multi-purpose carrier terminal would then take up two adjacent positions within this allocation scheme. To ensure full utilization of this frequency band so that no channel space remains unassigned, a comparatively large number of signaling channels would have to be superimposed on the speech channel. Even if there were a requirement for so large a number, another reason would discourage the use of such multi-purpose equipment. The distribution of the total send power over the individual subchannels would be rather unfavorable with both these transmission methods, so that a sacrifice would have to be made either in transmission range or in spectrum utilization. This is why most of all multi-purpose carrier terminals operate on the single-sideband transmission principle either in a 4 kHz allocation scheme, if initial plans already provide for the use of such multi-purpose terminals, or in a 5 kHz allocation scheme, if 2.5 kHz single-purpose terminals are also to be operated within the same allocation scheme.

The basic circuitry of a multi-purpose equipment corresponds to that of a single-purpose equipment (Fig. 51). The speech band and the super-imposed signaling channels are combined as they are translated from the voice frequency position to the intermediate frequency position in the send terminal, and segregated again in the receive terminal. The signaling channel units are in most cases incorporated in the carrier terminal.

The superimposed signaling channels may be amplitude-modulated or frequency-modulated channel units. Depending on whether the channels are intended for telemetering, remote control, or network protection applications, a channel spacing of 120 Hz or 480 Hz will be chosen. A combination of superimposed signaling channels of different bandwidth involves no difficulties (Fig. 58).

Fig. 58. Distribution of frequencies over the individual subchannels of a multi-purpose carrier terminal

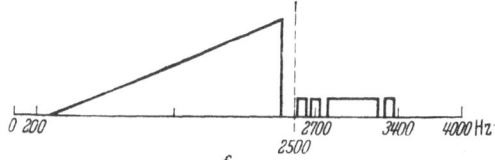

7.3 Multi-Purpose Carrier Terminals for Alternate Transmission

Independent of which transmission method is adopted, all carrier frequency terminals can be arranged so as to permit alternate operating options. The legitimate purpose of such a carrier terminal is always the transmission of speech. The alternate operating option offers the possibility of transmitting, without any increase in the bandwidth required, also signaling channels in such a manner that speech is interrupted while information is relayed over these channels. Obviously, this can be done only with very brief signals to avoid an unreasonable degradation of telephone communication. This precondition is satisfied only in the case of line protection signals and signals for direct tripping. Alternate use of a carrier terminal for telephony and direct breaker tripping is impracticable (see page 130). Therefore only such alternately operated equipment is of practical interest as is designed for telephony and network protection, all the more so as both-way transmission is required in either case.

The basic circuitry of alternately operated equipment corresponds to that of a single-purpose equipment. At the point of transition from the voice frequency stage to the intermediate frequency stage (single-sideband operation) or to the high-frequency stage (double-sideband operation or frequency modulation) an additional switch, an electronic switch in most cases, is provided at the place where an additional separating filter is provided in simultaneously operated multi-purpose terminals. Under the control of the external line-protection relays, this switch effects the change-over from speech to signal transmission and vice versa. This applies to the transmitting and to the receiving branches of the carrier terminal.

The alternate use of carrier terminals in such a manner that change-over devices outside the carrier terminal make the speech channel available to supervision and control signal or telegraph signal transmission for an extended period of time does not involve any special measures, as far as the transmission process is concerned. This is an infrequent application which is found only in telephone districts having a low call rate, so that measurands remote control signals or telegraph signals can temporarily be transmitted instead of speech.

7.4 Multi-Channel Carrier Terminals

For reasons of frequency conservation, the large majority of multi-channel carrier terminals operates with single-sideband transmission. If such terminals are intended for supervision and control signal transmission, they are specially adapted to this application. For use as telephone terminals (see page 86), they are borrowed, wherever possible, from postal systems and specifically adapted to operation over power lines. Where the frequency bands for the power network are distributed on the basis of a 4 kHz allocation scheme, less adaptation work is required than for a 2.5 kHz allocation, because the equipment used by postal administrations is invariably designed for operation in a 4 kHz allocation scheme. For use on power lines, however, the 4 kHz channel of a multi-channel carrier terminal would have to be utilized for speech and superimposed signaling channels if, compared with a 2.5 kHz allocation scheme, an unnecessary loss of frequency space, caused by extending the speech to 4 kHz without any reasonable benefit, is to be avoided. The idea of using one of the speech channels for supervision and control signal transmission might suggest itself as this would avoid the conversion of the

4 kHz channels into 2.5 kHz channels. The multi-channel carrier terminals may therefore be said to pave the way for the intrusion of the
4 kHz channels in power line carrier communication, although these are
here neither necessary nor even desirable. To avoid this, special 2-channel
telephone terminals have been developed for the 2.5 kHz allocation
scheme which, with the aid of simple accessories, can be expanded to
obtain 4-channel or 6-channel terminals.

The basic circuitry of a multi-channel carrier telephone terminal need
not be discussed here. One or a number of conversion stages are employed
to translate the voice frequency currents of each telephone circuit to
the HF position. For operation over power lines, it will be necessary to
improve the comparatively unsatisfactory overall attenuation characteristic within this rather broad transmission band by adapting the line
equipment accordingly. More than a single pilot channel should be made
available, if possible, for automatic volume control because, considering
the broad frequency bands, common line loss compensation should be
definitely limited to three telephone circuits. A separate ringing tone
channel is required for each telephone circuit.

Many years of experience have shown that the difficulties opposing
the introduction of multi-channel carrier telephone terminals are largely
due to the close ramification of power networks in the highly industrialized countries. Their elimination involves the necessity of providing
costly line equipment and of adapting postal carrier terminals to operation over power lines. There is yet another trouble. Power lines cannot
be readily disconnected for carrying out the necessary measurements
and installation work, a precondition for the use of multi-channel carrier
telephone terminals.

In less industrialized countries these difficulties do not exist, at least
not to the same extent. Moreover, the transmission terminals are here
used for a different purpose. The communication circuits are to be made
available for the public telephone service in the first place, and only a
small share is intended for power system operation. In spite of this,
multi-channel terminals affording only a small number of telephone
circuits are given preference also in these cases. The reasons are here
ease of installation, adjustment and maintenance and a greater flexibility
in network design. Moreover, the send power of each channel must be
sufficiently high, if a send amplifier common to several channels is used.

7.5 Survey of Equipment Models

During the early years of development the carrier terminal was felt
to be more or less of an outsider. It had to be set up separately.
The carrier section and the associated automatic switching unit were

accommodated in a metal cabinet. Since line tuning units were unknown at the time, the leads to the coupling capacitor had to be short and the cabinet was therefore set up in a corridor, in a cable room, or in another equipment room. Only the telephone sets were installed in the switching gallery. The metal cabinet is still frequently used as a means of accommodating a carrier terminal.

Rack-mounting versions have been used for many years in postal and railroad networks in line with the construction principles adopted for automatic dial exchanges. There is no objection to this because dust-free rooms are here available for the communication equipment and minimum space requirements are frequently an important consideration.

Some manufacturers decided at a very early stage to adopt the rack design also for power line carrier equipment. This can be done only if covers are provided to protect the individual constructional units from dust.

The curcuit elements, such as resistors, capacitors, transformers, and tubes have for a long time been mounted on vertical plates to facilitate trouble isolation up to the defective component. A horizontal arrangement was also used for these plates, which could then be pulled out of the cabinet like a drawer to permit a better utilization of the available space. The drawers had to be connected into the cabinet harness by means of extension cords.

For several years more and more nodal points have been set up in the communication networks of the power companies. They accommodate many carrier terminals, switching facilities and telemetering terminals. Special rooms have been made available for this purpose, dust-free and, in many cases, air-conditioned rooms. Disregarding the numerous earlier-type terminals which are still in use, i. e. terminals accommodated in metal cabinets and equipped with tubes, these communication rooms contain only rack rows which are accessible from both sides. Transistorization and modular design have reduced the size of the carrier terminals so that they now require only a quarter of the space occupied by the tube versions or even less than that. When existing communication systems have to be expanded, the room originally intended for the far larger tube-equipped carrier terminals will frequently suffice to house the new equipment. It is only a matter of replacing the old tube-equipped terminals by the smaller transistorized terminals which take up only a few shelves in a rack or cabinet. This saves space for expansion and avoids the cost of redesigning existing rooms or of providing new rooms.

In many countries carrier terminals of earlier design are still in operation. In some cases, however, the original housing conditions exist also for new terminals, i. e. these terminals are set up as a single terminal in a HT station. They are intended to provide both the carrier frequency and the switching functions. In either case cabinet-enclosed equipments are used. These individual terminals, distributed over the entire communication network, outnumber in most cases the terminals set up in communication centers.

Even today attempts are made to satisfy both requirements. The design objective is a standardized uniform version both for the outlying stations and the large centers. To achieve this purpose, the functional units of a carrier terminal must be designed to fit, in any optional arrangement, into standardized cabinets or standardized racks. This unitized construction principle permits all constructional differences between the various types and categories of equipment to be eliminated so that a uniform construction principle can be adopted for single-sideband, double-sideband, and frequency-modulated equipment on the one- and single-purpose, multi-purpose, and alternately operated equipment on the other hand. A maximum of adaptation to the design principles adopted for normal carrier practice and automatic dialing systems is highly desirable, to obtain a uniform appearance of the communication room facilities and, above all, assure low-cost production of standardized mechanical parts. In view of the differing design concepts of the various manufacturers, this ultimate aim will however never be fully realized.

Power companies frequently decided to replace their old communication equipment only when it was too late. The result was a heterogeneous conglomeration of all types of equipment in one room. Improvements in system management and closer control of the generation and distribution of power as well as the expansion of power networks involve an increased demand for more and better communication equipment. Old equipment will therefore be replaced sooner than in the past and, if they are still operational, the facilities thus freed are used as individual terminials on the periphery of the network.

Two points, among others, are of importance for the design of carrier frequency terminals: heat dissipation and measuring devices for fault isolation.

In the earlier tube equipment care had to be taken to remove the heat generated by the tubes inside the cabinet. In the present transistorized equipment heat dissipation is still a problem because transistors are rather susceptible to temperature variations and because the great packing density may result in an undesirable accumulation of heat, unless the necessary ventilation is included in the design.

The built-in measuring devices are required chiefly for lining up a new connection and later on for supervising the operability of the equipment. In the case of a carrier terminal set up in a remote station, substation operators having no particular training in communication engineering are enabled to isolate a trouble by means of some sort of remote diagnosis if a connection is in trouble. A minor fault may then frequently be cleared by some simple measures. A carrier communication specialist from the nearest nodal station will then be required only if the trouble is of a more complicated nature.

7.6 Power Supply

The power line carrier terminals set up either individually or in groups in the various stations of the power network, are fed from the station AC mains, if there terminals are still equipped with tubes, and from the 110 V or 220 V station battery provided for the auxiliary circuits of the power equipment. A number of different supply voltages must be generated inside the equipment for the heater, grid, and anode circuits as well as for the built-in automatic switching units and for the alarm circuits. To avoid the necessity of providing different batteries for feeding all these circuits, the necessary supply voltages are derived from the AC mains by way of a built-in power supply unit. It is true that large communication systems are always equipped with a battery of, say, 48 V or 60 V for feeding the automatic dial exchange and other communication equipment. This battery is however unsuitable for direct feeding of the consumer circuits of a carrier terminal equipped with vacuum tubes, because various different and partly also higher voltages (anode voltage) are required.

In power generating stations a dependable 50 Hz power supply is in most cases ensured. Adequate protection against a mains voltage outage in transformer sub-stations exists only if the station is fed from several points.

The communication equipment is obviously required most any time a fault occurs in the power network. The trouble source must be traced and action must be taken to eliminate the fault. If the AC mains is out of order as a result of such a fault, a stand-by power supply of 220 V, 50 Hz must take over the supply of the carrier communication equipment. Large communication centers are equipped with a gasoline engine- or a Diesel engine-driven AC generator which is cut in automatically in the event of a power outage, in particular where other circuits, in addition to the communication equipment, must be fed. If the power requirements are small, say, up to 5 kVA, two-unit or three-unit motor generator sets are used. These are driven from the station battery. What is important is that no interruption will in either case occur in the power supply to the connected communication equipment, if a short interruption of the communication channels is not permissible as in the case of carrier channels for network protection. The continuously running engine- or motor-driven generator sets are a rather heavy load on the budget, all the more so as a spare unit must be available to allow for periodic maintenance and general overhaul.

In those cases where a brief interruption in the power supply to the communication equipment is permissible between outage and restoration of the AC mains supply, a simple emergency power supply will also serve the purpose. The equipment used may be motor-generator sets or rotary converters, which are automatically cut in and cut out in dependence on the AC mains voltage.

The length of a no-supply period corresponds to the run-up time of the motor. Difficulties will hardly arise if the equipment to be fed is used for speech or telemetering signal transmission.

The border between power supply units with and without inter-ruption of the channels becomes blurred more and more as suitable methods have been devised to reduce the no-supply period and its effect on the communication equipment. Modern emergency supply units are designed to run at a high speed and have light-weight rotors so that full speed is attained within fractions of a second even at a small input power. Reservoir capacitors are provided in addition, to avoid a decrease in the voltage during this brief run-up time.

The costs of the equipment required to ensure power supply continu-ity for tube equipments are felt to be a heavy burden in particular in those cases where they cannot be prorated to a large number of carrier terminals. With a single carrier terminal in a medium-voltage station, which may not even have a 110 V station battery, the costs of the battery and the stand-by power supply unit would exceed those of the carrier terminal. In most cases a lower capacity will do. A "transistorized converter" may be adequate. It converts the DC voltage into an AC voltage without the need for tubes or mechanical and moving parts.

Present-day carrier terminals are equipped with transistors instead of tubes so that no heater power is required. The energy requirement is much lower, all the more so as the automatic switching units used for setting up calls to the nodal stations are self-contained units which can be fed directly from an exchange battery. Nor are the high voltages re-quired as were formerly used for the tubes of earlier models (anode voltages). All carrier terminals can therefore be connected to a communi-cation equipment battery without the need for extra facilities to safe-guard the power supply. A power outage has no consequences as regards carrier communication. One exception to this rule are send amplifiers of a particularly high rating as are occasionally required to span very great distances. These amplifiers are still equipped with tubes because tran-sistor circuits providing so high an output are too expensive. These send amplifiers are fed from the battery by way of static DC–AC converters, i. e. they require supplementary facilities for their power supply.

In transistorized carrier terminals, however, all problems related to grounding and DC isolation of certain circuits within and without the equipments require a far more careful examination than in tube equipments because transistor circuits are DC coupled and because the transistors are very susceptible to overvoltages.

8. Measuring Methods and Measuring Equipment

During the first ten years of development up to about 1930 the standards used for assessing the quality of a power line carrier communication circuit were essentially of a subjective nature. Objective measurements were exceptional cases. A frequency meter of the variety used in radio transmission systems was relied upon to adjust the oscillators as accurately as possible and to tune the resonant traps. The carrier output of the transmitter was measured in the line circuit by means of a hot-wire instrument. At the receiving end, the incoming carrier voltage was measured with a moving coil instrument after rectification in a vacuum tube. This method offered some assurance that the carrier frequency was properly transmitted and it gave an approximate picture of signal attenuation in the carrier channel. Vacuum tube voltmeters were used to measure the HF voltages in the receivers and in by-pass circuits having no telephone stations connected to them.

These measurements covered only double-sideband transmissions and, more than anything else, were actually meant to provide a certain assurance of dependable signal transmission. For an objective judgment of the quality of speech transmission the instruments available at the time were not sophisticated enough. The measuring instruments installed in present-day carrier terminals for checking the characteristic operating values (HF send current and send voltage, receive current, auxiliary voltages for vacuum tubes and automatic switching unit, etc.) are equally unsuited for this purpose. Recourse must be had to more involved measuring facilities to be able to determine the attenuation of a circuit as a fonction of the frequency and to keep tab on the level conditions along the line, or to measure the noise level both in the HF position and in the VF position. Special measuring equipment is also required for checking the dial pulses.

The advances scored in the carrier communication technique, in particular the introduction of single-sideband transmission, involved the necessity of more extensive and more complicated measurements. New measuring equipment, tailored to the specific requirements arising in connection with carrier communication, had to be developed. Where new communication systems are cut over to service or where extensive modification work becomes necessary in existing networks, there is a need for comprehensive measuring facilities to eliminate right from the outset any weaknesses which could later be traced back to inadequacies in the verification of line characteristics and the resultant poor adjustment of the transmission equipment. However, for maintaining a properly lined-up communication system in satisfactory operating condition,

Measuring equipment	Application
Voice frequency measuring set	Measurements within the VF range from 300 Hz to 4000 Hz. Transmission of standard level; measurement of level, attenuation, and gain; measurement of impedance
Carrier frequency test set	Measurements in the HF range from 10 kHz to 500 kHz. Transmission of test level; measurement of level, attenuation, and gain; measurement of impedance
Pulse recorder	Checking dial switches and pulsing relays for proper contact make; recording of dial pulses
Noise measuring set	Location of HF interference sources

Fig. 59. Measuring equipment for maintenance purposes

these measuring facilities would be too expensive and, after all, unnecessary. Maintainers can get along with a limited number of measuring devices. These are used for periodic checks which are intended to detect incipient failures.

The measuring equipment required for this purpose and its application have been tabulated in Fig. 59.

The impedance of a wave trap, with or without tuning set, can best be determined by a current-voltage measurement with the aid of a carrier frequency test set (Fig. 60).

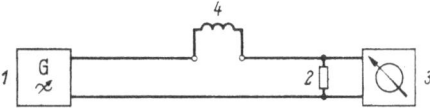

Fig. 60. Trap testing circuit.
1 Carrier frequency oscillator with accurate frequency calibration; *2* Measuring resistor, approx. 10 Ω; *3* Level meter; *4* Wave trap

Adverse weather conditions may affect wave traps or their tuning media in the course of time. Checks can be carried out only with the line disconnected and grounded (Fig. 61).

During line-up operations, sensitivity, equalization, and VF levels of the carrier terminals are adjusted to the local conditions. These preparations include the adaptation of the built-in power supply unit to the mean supply voltage, because most carrier terminals are unable to accommodate auxiliary voltage fluctuations exceeding ± 10% of the rated value.

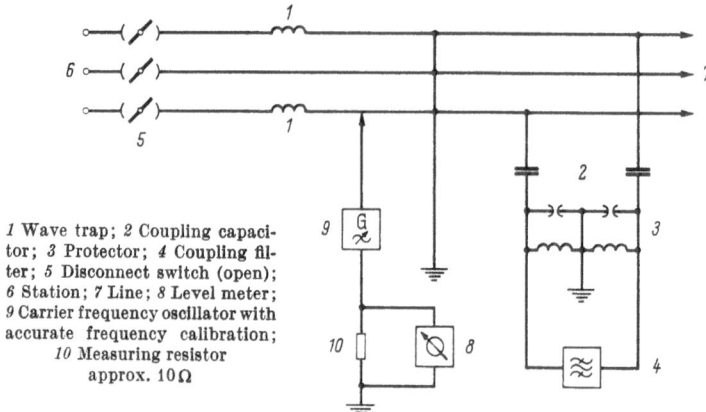

1 Wave trap; 2 Coupling capacitor; 3 Protector; 4 Coupling filter; 5 Disconnect switch (open); 6 Station; 7 Line; 8 Level meter; 9 Carrier frequency oscillator with accurate frequency calibration; 10 Measuring resistor approx. 10 Ω

Fig. 61. Circuit for checking wave traps in the switchyard.

Line-up measurements can only partly be performed with the aid of the built-in measuring instruments. Measuring equipment for checking the level at the VF terminals, for instance, is one of the items missing in most equipment models because of its high cost. Line-up includes also a check of those operating values which are independent of local conditions and have been permanently adjusted in the factory. This re-check is made to discover any damage which may have occurred during transport.

With the carrier terminal checked and adjusted to local conditions, the overall attenuation and its frequency response are measured in the VF position. The frequency response should stay within the limits recommended by the CCITT for an international two-wire circuit.

The transmission conditions on a power line are sometimes difficult to analyse if, for instance, some of the spur lines in medium-voltage networks are to be left untrapped to save the high costs of the wave traps. Planning work must here be preceded by measurements to investigate all possible ways in which a given transmission problem can best be solved. If the omission of traps is likely to result in an impermissible attenuation under the different switching conditions of the power system, measurements will become necessary to determine the carrier frequencies at which acceptable performance is still possible (Appendix 9.1).

Measurements become a necessity in power lines where the frequency response of the characteristic impedance is not known. This applies to lines having short power cable sections interposed between overhead wire sections. Further, it will be advisable to carry out measurements prior to the installation of the communication system where attenuation conditions are critical because of frequent hoarfrost, where unusual noise levels must be anticipated, and where re-use of certain carrier frequencies is desirable in view of the existing spectrum crowding. All these tests are mostly performed with the power line under load, partly because users cannot be expected to put up with a power interruption, partly because the factors to be determined – the noise level for instance – are a function of the power voltage.

Measurements made in connection with equipment line-up and periodic maintenance fall into the category of operational measurements. For system planning purposes, however, the measurements made directly on the power line are of particular interest. These are intended to determine the impedance, the attenuation, and the noise level. Different methods and different types of equipment are used for such measurements. To achieve the desired uniformity and to arrive at readily comparable results, the CIGRE has issued a set of suitable recommendations [13].

8.1 Measuring Methods

Where the emphasis is on an extremely great accuracy, bridge circuits are normally adopted for impedance measurements. A simpler method permits the complex impedance of a power line to be evaluated with sufficient accuracy by means of current and voltage measurements. Being less cumbersome and time-consuming, this method is used to good advantage on lines affected by high noise voltages, as it eliminates the need of adjusting reference resistors for minimum instrument deflection. The equipment is the same as is used for attenuation measurements, that is, a level oscillator and a selective level meter.

a) Magnitude of Impedance

In the case of $R < X$, $|X| = R \, 10^{\frac{p_1 - p_2}{20}}$

Example calculation: $R = 1.0\,\Omega$. Measured level $p_1 = +\,8.7$ db,
$$p_2 = -\,35 \text{ db.}$$

$$|X| = 1.0 \times 10^{\frac{43.7}{20}} = 150\,\Omega$$

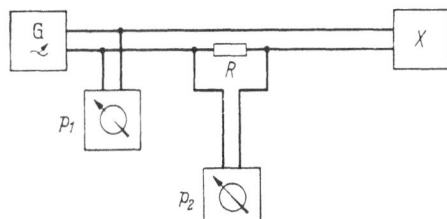

Fig. 62. Circuit arrangement for measuring the magnitude of impedances (For symbols see Fig. 61)

b) Real and Imaginary Components

$$X = A + jB$$

1. Measurement without capacitor (pushbutton T operated) yields the value $|X_1|$.
2. Measurement with capacitor yields the value $|X_2|$.

Calculation of A and B from X_1 and X_2:

$$B = \frac{X_1^2 - X_2^2 + X_c^2}{2\,X_c}; \qquad A = \sqrt{X_1^2 - B^2}\,,$$

where $X_c = \dfrac{1}{\omega C}$.

Example calculation $|X_1| = 140\,\Omega$, $|X_2| = 100\,\Omega$.
$$C = 16000\,\text{pF};$$
$$f = 100\,\text{kHz};$$
$$\frac{1}{\omega C} = 100\,\Omega$$

$$B = \frac{(1.96 - 1 + 1) \times 10^4}{2 \times 10^2} = +\,97\,\Omega\;(\text{induktive})$$

$$A = 10^2 \sqrt{1.96 - 0.98^2} = 100\,\Omega$$

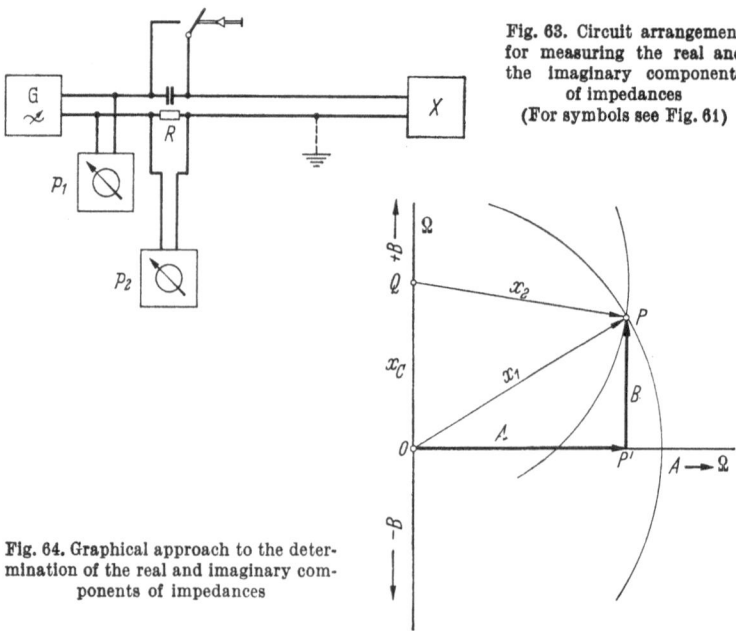

Fig. 63. Circuit arrangement for measuring the real and the imaginary components of impedances (For symbols see Fig. 61)

Fig. 64. Graphical approach to the determination of the real and imaginary components of impedances

Graphical approach for A and B (Fig. 64):

Draw a circle with radius $|X_1|$ about the center point 0. Plot $X_c = 1/\omega C$ on the ordinate, starting at point 0, and ending at point Q.

Draw a circle with radius $|X_2|$ about point Q. This circle intersects the circle with radius $|X_1|$ at point P. P gives the impedance to be determined and $OP' = A$, $PP' = B$.

For series measurements it will be convenient to use previously prepared diagrams.

8.2 Measurements on a De-Energized Power Line without Line Equipment

With the power source disconnected, the characteristic impedance and the attenuation can be measured directly on the line. Overvoltage arresters and grounding reactors should be provided to guard against the effects of atmospheric discharges and voltages induced by adjacent power lines. The grounding coils should be rated for an inductance of $L = 50$ mH and for a natural resonance $f_0 \geqq 200$ kHz. Above this fundamental resonance there must be no further resonance (short circuit resonance) in the range up to 500 kHz. With excessive induction from neighboring lines (lines on the same tower structure) it may become necessary to protect the level oscillator and the level meter with the aid of capacitors. The applicable safety regulations must be adhered to. The power line must be grounded while the measuring arrangement is being connected and each time this arrangement is modified.

a) Characteristic Impedance

The input impedance of a homogeneous line not terminated in its characteristic impedance can be represented by a frequency-dependent curve showing real maximum and minimum values X_{max} and X_{min}. The spacing b between two minimum or two maximum values depends on the length l of the line and amounts to

$$b = \frac{c}{2l}$$

where $c = 186\,000$ miles/s. (velocity of propagation of electromagnetic waves).

At a line length of $l = 60$ miles, the spacing $b = 1.55$ kHz. Assuming the usual line lengths, the characteristic impedance can be determined with fair accuracy from the consecutive maximum and minimum values of the impedance, using the formula

$$Z = \sqrt{X_{max} X_{min}} \,.$$

Determination of X_{max} and X_{min} is based on the method indicated for impedance measurements. With phase-to-phase coupling and a line of sufficient length (Fig. 65), the line may be shorted rather than terminated in R_E in station E, while measurements are carried out at station D. In the case of short lines it will be advisable to provide a terminating resistance R_E approximating the expected characteristic impedance of

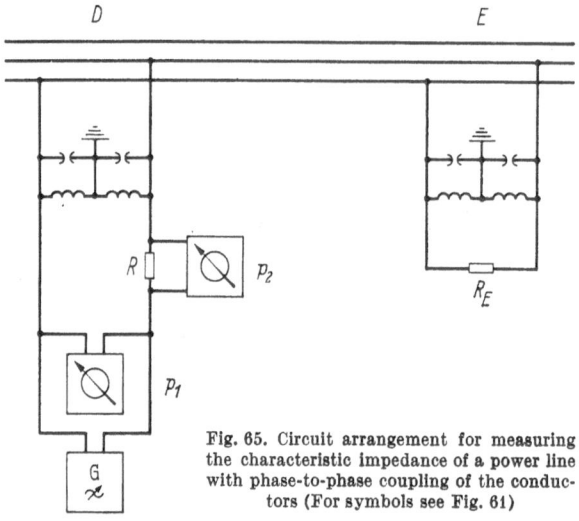

Fig. 65. Circuit arrangement for measuring the characteristic impedance of a power line with phase-to-phase coupling of the conductors (For symbols see Fig. 61)

the line because, with a low line attenuation and a shorted (or open) line, X_{max} will take on too high a value and X_{min} too a low value, and these values may not be easily measured.

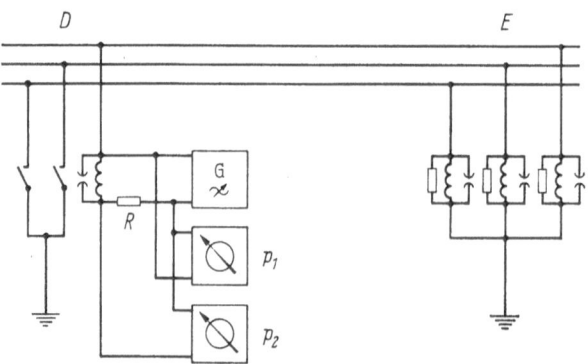

Fig. 66. Circuit arrangement for measuring the characteristic impedance of a power line with phase-to-ground coupling of the conductors (For symbols see Fig. 61)

If the limit values of the input impedance (see page 50) are to be measured as characteristic impedance values, a phase-to-ground coupled line must be terminated at E, unless the line under test is a very long one (Fig. 66). With an appreciable reaction from the output to the input, the geometric mean of the input impedance would show an excessive deviation from the limit value because of the losses in the station and at the beginning of the line. If each of the three conductors is terminated in an impedance of approximately 400 Ω against ground, the reaction from the end of the line can be held well within reasonable limits. An open-circuit or a short-circuit condition at the beginning of the non-coupled conductors will produce the maximum and the minimum values, respectively, of the characteristic impedance which may occur with station impedances of different magnitude.

b) Attenuation

For line attenuation measurements the impedances R_i at the beginning and at the end of the line are made equal to the measured characteristic impedance Z (for phase-to-phase coupling according to Fig. 67, for phase-to-ground coupling according to Fig. 68).

The attenuation is then

$$a = p_1 - p_2 - 6 \text{ db.}$$

Most power lines cannot be regarded as being completely homogeneous and minor frequency-dependent periodic variations must be ex-

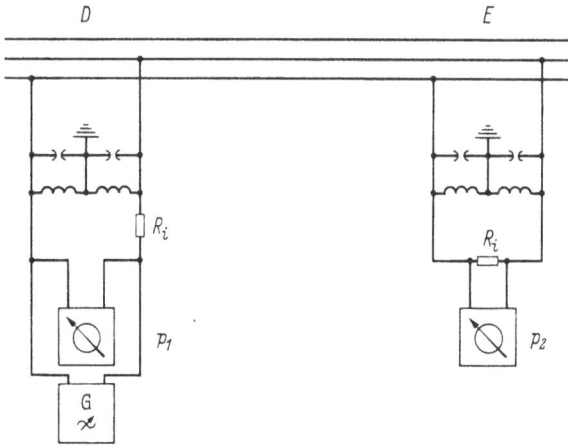

Fig. 67. Circuit arrangement for measuring the attenuation of a power line with phase-to-phase coupling of the conductors (For symbols see Fig. 61)

pected, even with phase-to-phase coupling of the conductors, both during attenuation and characteristic impedance measurements. In the phase-to-ground coupling case, these variations will always be present in view of the imperfect termination of the line. Mismatch points along the right-of-way of the power line, such as spurs and branch lines, may result in a greater frequency response. This might involve the necessity of terminating the line section used for carrier communication in different impedances for optimum matching.

The attenuation is then

$$a = p_1 - p_2 - 6\,\mathrm{db} + 10\log\frac{Z_2}{Z_1}\,\mathrm{db}\,.$$

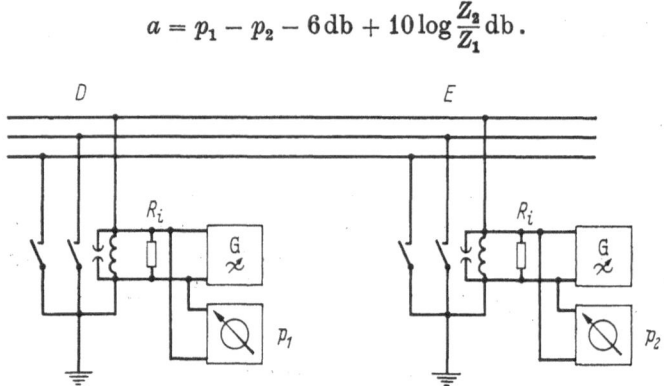

Fig. 68. Circuit arrangement for measuring the attenuation of a power line in the phase-to-ground coupling case (For symbols see Fig. 61)

8.3 Measurements on an Energized Line

Line tuning units of the broad-band variety are used for power line coupling. The characteristic impedance of these filter circuits is not constant over the entire passband range. If a HF entry cable is employed, a frequency-dependent complex impedance will be measured at its input if the cable is terminated by the coupling circuit and the power line is connected.

a) Impedance and Attenuation

The input impedance and the attenuation of the coupling circuit can be measured separately by replacing the coupling capacitors by suitable capacitors of equal capacitance and simulating the line by an ohmic resistance equaling the magnitude of the characteristic impedance. In most cases it will suffice here to determine Z by calculation (see page 49).

The input impedances of coupling circuits can be measured by employing the method indicated in Fig. 69.

The attenuation can be measured in another circuit arrangement (Fig. 70) with R_i representing the internal resistance of the signal transmitter.

The attenuation of coupling circuits is found from the formula

$$a_{cpl} = p_1 - p_2 + 10 \log \frac{Z}{R_t} - 6 \, \mathrm{db}.$$

The input impedance of a coupling circuit, with the line connected, can be determined with a similar circuit arrangement (Fig. 69). Using a

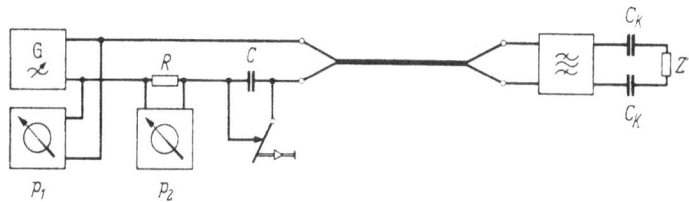

Fig. 69. Circuit arrangement for measuring the input impedance of a coupling circuit (For symbols see Figs. 53 and 61)

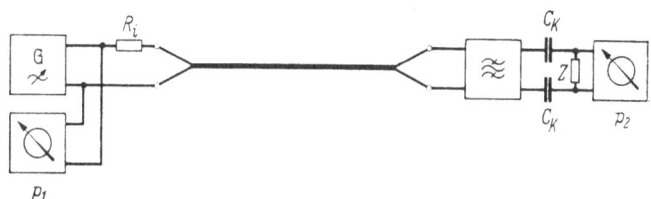

Fig. 70. Circuit arrangement for measuring the attenuation of a coupling circuit (For symbols see Figs. 53 and 61)

suitable measuring arrangement (Fig. 71), the transmission loss of the total circuit will be

$$a_t = p_1 - p_2 + 10 \log \frac{R_{t2}}{R_{t1}} - 6 \, \mathrm{db}.$$

This measurement covers also the shunt loss introduced by the stations and, using the appropriate circuit for phase-to-ground coupling, the incremental attenuation with phase-to-ground coupling.

b) Noise Level

The noise level is measured with the level meter connected to the line by way of the coupling circuit (Fig. 71, station E).

It is customary to refer the measured noise level introduced by power transmission to a bandwith of 2.5 kHz using the following formula:

$$p_5 = p_x + 10 \log \frac{2.5}{x} + a_{cpl} + 10 \log \frac{Z}{R_i}.$$

where

$$x \quad = \text{bandwidth of level meter}$$
$$a_{cpl} = \text{attenuation of coupling circuit.}$$

Eliminating the bandwidth correlation factor, this expression can be used for calculating the level of a selective interferer.

The level oscillator required need not have the extremely great frequency accuracy sometimes necessary for special measurements on communication circuits. The level meter should have a good selectivity

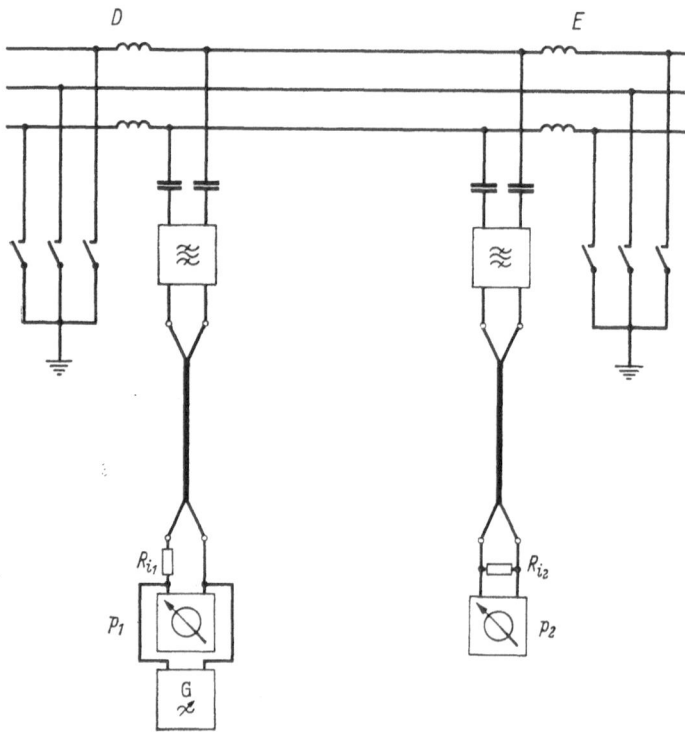

Fig. 71. Circuit arrangement for measuring the transmission loss of the total circuit (For symbols see Figs. 53 and 61)

and a high harmonic distortion attenuation to permit also measurement of low signal levels in the presence of a high noise level. Noise levels should be determined through the measurement of an approximate rms value.

In some cases the selectivity of the receiver is not desired. Since broad-band measurements are easier and less time-consuming, it will be an advantage if the level meter can be switched over from selective measurements to broad-band measurements.

The measuring equipment used should possess characteristics similar to those stated below (see page 149).

a) *Level oscillator*

Frequency continuously variable	10 kHz to 500 kHz
Frequency error	$< 0.5\%$
Incremental fine adjustment	± 5 kHz
Fine adjustment error	< 100 Hz
Internal resistance	zero Ω (< 3 ohms)
Output voltage level across $R_a = 75\ \Omega$	$+10$ db
Output voltage level continuously adjustable	0 to $+10$ db

b) *Selective level meter*

Frequency range	10 kHz to 400 kHz
Measuring range, selective	-70 db to $+20$ db
Measuring range, broad-band	-50 db to $+20$ db
Passband range, selective	$f_0 \pm 90$ Hz
Stop-band attenuation, selective	> 60 db at $f_0 \pm 300$ Hz
Non-linear distortion attenuation	> 70 db
Input impedance	> 5 kΩ

9. Appendix

9.1 Effect of Spur Lines on Carrier Communication Circuits

AC currents in open-wire lines are propagated at a speed equaling almost the velocity of light waves $c = 300000$ km/s. (186000 miles/s.) This applies both to the power current frequencies of $16^2/_3$ Hz, 50 Hz and to the carrier frequencies of 15 kHz to 500 kHz as used in power line communication systems. The frequency f refers to the number of oscillations per second, from which the "wave length" can be determined as

$$\lambda = \frac{c}{f} = \frac{300\,000\,000}{f_{\text{Hz}}} \text{ meters} .$$

The power current frequencies of $16^2/_3$ Hz and 50 Hz therefore correspond to the wave lengths 18000 km and 6000 km while the carrier frequencies 50 kHz and 300 kHz correspond to the wave lengths 6000 meters and 1000 meters.

The fact that lines branched off from the main power line (spur lines in particular) may result in an undesirable loss of HF energy was already realized in the early days of power line carrier communication. It was even found that the HF energy will be completely canceled if a certain ratio happens to exist between the length of the spur line and the length of the electric wave used for signal transmission. Under certain switching conditions, a spur line may therefore mean a short circuit for certain frequencies at the branch-off point. On the other hand, it was found that a loop of suitable dimensions in this spur line would raise the impedance of the spur line so that an appreciable loss in carrier energy will not occur. This can be explained as follows:

Using the symbols

W_{1o} to denote the input impedance of the open spur line,
W_{1sh} for the input impedance of the shorted spur line,
l for the length of the spur line,
Z for the characteristic impedance of the line,

then

$$W_{1o} = -jZ\cot\frac{2\pi fl}{c} \quad \text{and} \quad W_{1sh} = jZ\tan\frac{2\pi fl}{c}.$$

From these two equations with $c/f = \lambda$ it is possible to determine the critical line lengths (Fig. 72).

Effect on carrier frequency	Switching condition of spur line	
	1. Open circuit $W_{1l} = -jZ\cot\frac{2\pi l}{\lambda}$	2. Short circuit $W_{1k} = jZ\tan\frac{2\pi l}{\lambda}$
A. $W = 0$ (short circuit)	if $\cot\frac{2\pi l}{\lambda} = 0$	if $\tan\frac{2\pi l}{\lambda} = 0$
This holds only for the following angles:		
	$\dfrac{\pi}{2}, \dfrac{3\pi}{2}, \dfrac{5\pi}{2}, \ldots$	$0, \pi, 2\pi, 3\pi, \ldots$
The expression $2\pi l/2$ takes on these angle values if		
	$\dfrac{l}{\lambda} = \dfrac{1}{4}, \dfrac{3}{4}, \dfrac{5}{4}, \ldots$	$\dfrac{l}{\lambda} = 0, \dfrac{1}{2}, \dfrac{2}{2}, \dfrac{3}{2}, \ldots$
Therefore, short circuit if		
	$l = \dfrac{1}{4}\lambda, \dfrac{3}{4}\lambda, \dfrac{5}{4}\lambda, \ldots$	$l = 0, \dfrac{1}{2}\lambda, \lambda, \dfrac{3}{2}\lambda, \ldots$
B. $W = \infty$ (trap)	if $\cot\frac{2\pi l}{\lambda} = \infty$	if $\tan\frac{2\pi l}{\lambda} = \infty$
This holds only for the following angles:		
	$0, \pi, 2\pi, 3\pi, \ldots$	$\dfrac{\pi}{2}, \dfrac{3\pi}{2}, \dfrac{5\pi}{2}, \ldots$
The expression $2\pi l/\lambda$ takes on these angle values if		
	$\dfrac{l}{\lambda} = 0, \dfrac{1}{2}, \dfrac{2}{2}, \dfrac{3}{2}, \ldots$	$\dfrac{l}{\lambda} = \dfrac{1}{4}, \dfrac{3}{4}, \dfrac{5}{4}, \ldots$
Therefore, trapping if		
	$l = 0, \dfrac{1}{2}\lambda, \lambda, \dfrac{3}{2}\lambda, \ldots$	$l = \dfrac{1}{4}\lambda, \dfrac{3}{4}\lambda, \dfrac{5}{4}\lambda, \ldots$

Fig. 72. Critical spur line lengths l as a function of the wave length λ of a carrier frequency circuit

Depending on their length and on the switching status at their ends, spur lines not intended for carrier transmission (Fig. 74a) introduce the following extreme conditions on the carrier circuit:

Case No.	Length	Switching condition at the end	Effect on carrier circuit
1 A	$\lambda \times \dfrac{1}{4};\ \dfrac{3}{4};\ \dfrac{5}{4}\ \cdots$	Open circuit	Short circuit
2 B		Grounded (short circuit)	Trap
2 A	$\lambda \times \dfrac{1}{2};\ \dfrac{2}{2};\ \dfrac{3}{2}\ \cdots$	Grounded (short circuit)	Short circuit
1 B		Open circuit	Trap

Fig. 73. Effect of spur lines on carrier frequency circuits

In the cases 1 A and 2 A it is imperative that the spur line be trapped immediately at the branch-off point to prevent cancellation of the carrier energy, whereas wave traps need not be provided in cases 1 B and 2 B. As can be seen, a critical case exists whenever the length of the

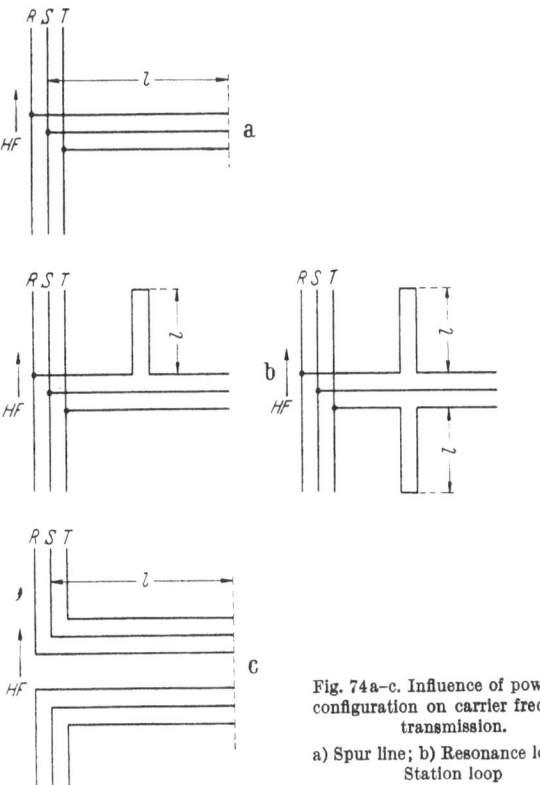

Fig. 74a–c. Influence of power line configuration on carrier frequency transmission.

a) Spur line; b) Resonance loop; c) Station loop

spur line corresponds to an even multiple of the quarter wave $(1/4\ \lambda)$. Arranging the wave trap at the end of the line, for the sake of convenience, is absolutely inadmissible in such a case.

Spur lines may as well be used as a part of the carrier communication network. Neither of the two extreme cases will then exist, because the line end is then connected to a carrier terminal.

Case 2 B would apply for rating a "resonance loop" (Fig. 74b), also called "antenna trap". Trapping could therefore be accomplished by inserting a loop at certain points of the line. However, to achieve a blocking effect in the frequency range from 15 kHz to 500 kHz, the length of the loop would have to range between 3 miles and 0.1 mile, depending on the carrier frequency to be trapped. To trap a number of carrier frequencies, several loops would be required at one point, and a change in the carrier frequencies would involve a corresponding change in the length of the power line loops. This trapping method has never been used in practical applications because of excessive space requirements, high cost, and lack of flexibility, in short, because the entire arrangement is too unwieldy.

The use of bundled conductors for overhead transmission lines has brought about a temporary revival of this subject. Two-conductor bundles were tentatively installed at the station entry of carrier frequency terminals, not for the purpose of energy transmission but as a conveniently constructed resonance loop. Nevertheless, the expenditure is comparatively high also in this case. Such a configuration may serve as a trap for a specific carrier frequency but it can hardly be designed so as to block a wider frequency range. Besides, the lengths of the conductors of a three-phase system exhibit considerably greater differences (Fig. 74b) than in the case of regular wave traps. As a result, difficulties in the operation of a network protection system are more likely to occur.

Another point of interest in system construction is the station loop (Fig. 74c). With phase-to-phase coupling, the station loop may be regarded as a resonance loop according to case 2 B, provided the go-line and the return-line are interconnected, which is usually the case. However, this is no longer possible where phase-to-phase coupling is employed. To avoid energy losses in carrier frequency transmission, the carrier path is through-connected by way of a by-pass circuit. In those cases, however, where the carrier circuit is to continue along the main route and, at the same time, is to be branched off so as to follow the path of the lead-in loop, the length of the latter may equal an even multiple of the quarter wave length without causing any undesirable effects. The carrier terminal represents a terminating resistance lying between the two extreme cases of short circuit and open circuit. Similarly, no difficulties will arise if two carrier circuits arriving from opposite directions on the main line are looped in to the station.

9.2 Effect of Wave Traps[1]

The impedance presented by the wave traps is practically insignificant as far as the power frequency is concerned, so that the voltage drop caused by the operating current can be safely neglected. For the carrier frequency currents, however, the impedance should be higher than or at least equal to the characteristic impedance of the line, if an adequate blocking effect is to be ensured.

If the power network station at the end of the line is disconnected and the line is not grounded, the station will not introduce any attenuation in the transmission circuit. If, on the other hand, the line is grounded or the station connected, the wave trap or, station and wave traps, are arranged in parallel with the coupling circuit. This condition produces an increase in attenuation which should be kept as low as possible by an adequate rating of the trap impedance. This "shunt loss" is a measure of the efficiency of the wave trap. Its magnitude depends not only on the inductance of the wave trap, but also on the tuning of the trap and on its location within the station circuitry.

To obtain a sufficiently low shunt loss, the reactance need not be particularly high (Fig. 75, curve b). To trap one or two frequency bands within the working range, traps of a low inductance, say 0.2 mH are

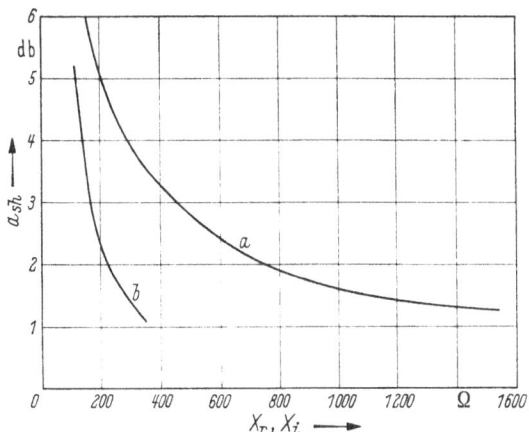

Fig. 75. Shunt loss with phase-to-phase coupling.
a As a function of an ohmic trap resistance; b As a function of an inductive trap reactance

used. They are combined with a capacitor so as to form a resonant circuit (Figs. 20a, c and 76). Where all frequencies within the available range have to be blocked, the traps must be rated for considerably higher

---

[1] According to D. FRANKE.

inductances. A sufficiently large reactance can be attained without any tuning operations if, for instance, an inductance of 2.0 mH is used to cover all frequencies above 35 kHz. An inductance of 1.0 mH will suffice for trapping all frequencies above 100 kHz.

Since a switching station represents normally a capacitance, the inductive reactance both of a resonated and a non-resonated wave trap may be partly, if not fully, compensated by the capacitive reactance of the station. A substantial increase in the shunt loss must therefore be expected if the short circuit or ground is removed and the line is connected

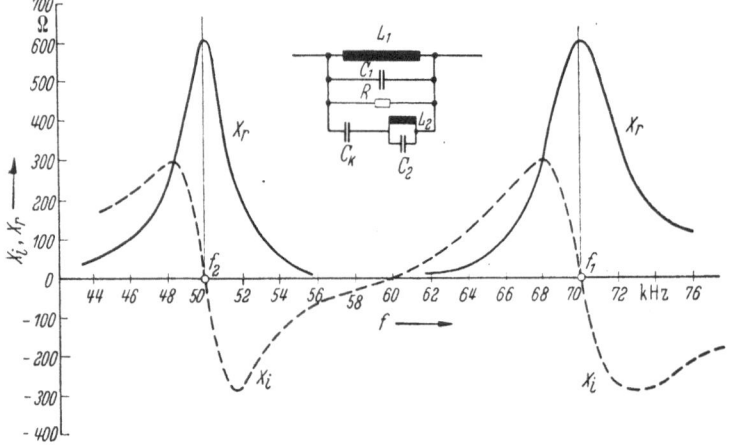

Fig. 76. Tuning a wave trap to two frequency bands
$L_1 = 0.2$ mH, $L_2 = 0.15$ mH, $R = 600$ Ω, $f_2 = 50$ kHz, $f_1 = 70$ kHz

to the station. Using 2-mH wave traps, a station capacitance of 5000 pF will permit transmission at a shunt loss of 1.8 db or less only at frequencies of ≥ 60 kHz. Assuming the same station capacitance, a 0.2-mH wave trap, resonated to a frequency of 50 kHz, will introduce a shunt loss of ≤ 1.8 db for the frequency band 48.5 kHz to 56.0 kHz.

Practical experience has shown that a compensation of the blocking impedance by the station capacitance is a very rare occurrence. A value of > 5000 pF may therefore be normally assumed for practical applications. The following applies in connection with smaller capacitance values:

Spur lines, either open-circuited or terminated by a switching station which has a small capacitive reactance, may take on any impedance, this being a function of the length of the line and of the carrier frequency. The stations proper may produce a similar effect under certain switching conditions, if the bus bar system has a considerable extension, as is the case in extra-high voltage systems. Depending on the frequency, the reactances may be capacitive or inductive, so that a certain portion of the frequency range to be blocked cannot be used for signal transmission.

The wave trap must therefore feature a sufficiently large ohmic resistance within the blocking range. Assuming the same shunt loss, the ohmic resistance component must be appreciably higher than the uncompensated reactance (Fig. 75, curve a). If $a = 1.8$ db, the reactance X_i will be 250 Ω and the ohmic resistance X_r 900 Ω.

The easiest way of obtaining a resistive component in the trap impedance is to connect a pure ohmic resistance in parallel with the coil. Although the blocking range of 0.2-mH traps, resonated to one frequency or to two frequencies, is thereby narrowed down, it will still suffice for

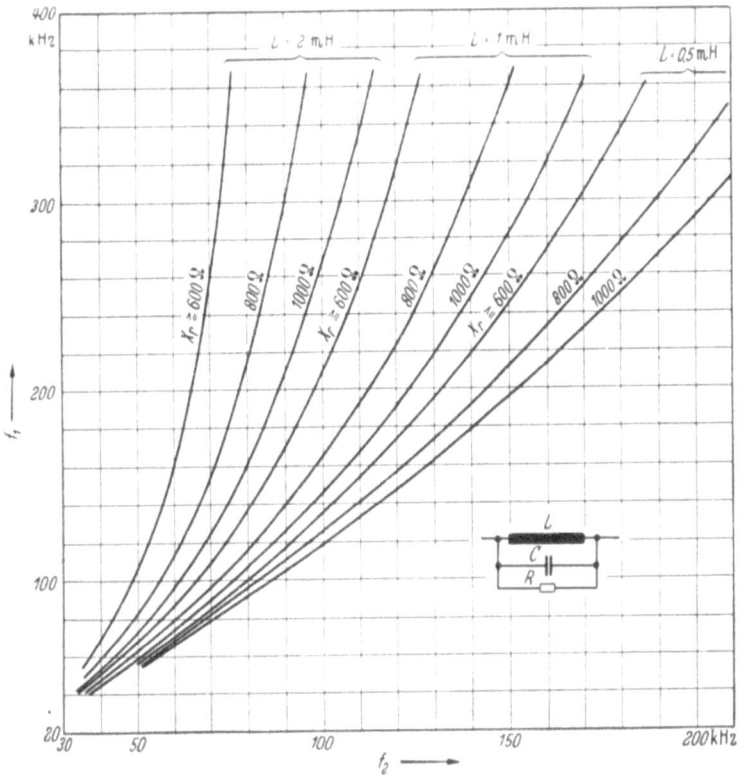

Fig. 77. Blocking range of broad-band traps as a function of inductance and minimum value of the ohmic component of the blocking impedance, circuit arrangement 1

(f_1 = upper edge frequency, f_2 = lower edge frequency of blocking range)

5 kHz bands in the region above 50 kHz with single-frequency tuning, and in the range above 90 kHz with two-frequency tuning, if a higher insertion loss (approximately 2.6 db) is acceptable.

This method cannot be adopted for non-resonated traps of 1.0 mH or 2.0 mH. To obtain a purely resistive component of ≥ 600 Ω also in these cases, a reactive component of 1200 Ω would be required. As a result, the blocking range of a 2.0-mH trap would begin only somewhere in the region of 100 kHz.

To get around this problem, a capacitor is connected in parallel with the coil, in addition to the ohmic resistance. It is then possible to determine, at a given inductance, the frequency range f_1 to f_2 within which the required minimum of the resistance X_r will still be ensured (Fig. 77). With a 2-mH trap, for instance, the value of the resistive component X_r will not fall below 800 Ω within the range from $f_1 = 300$ kHz to $f_2 = 90$ kHz.

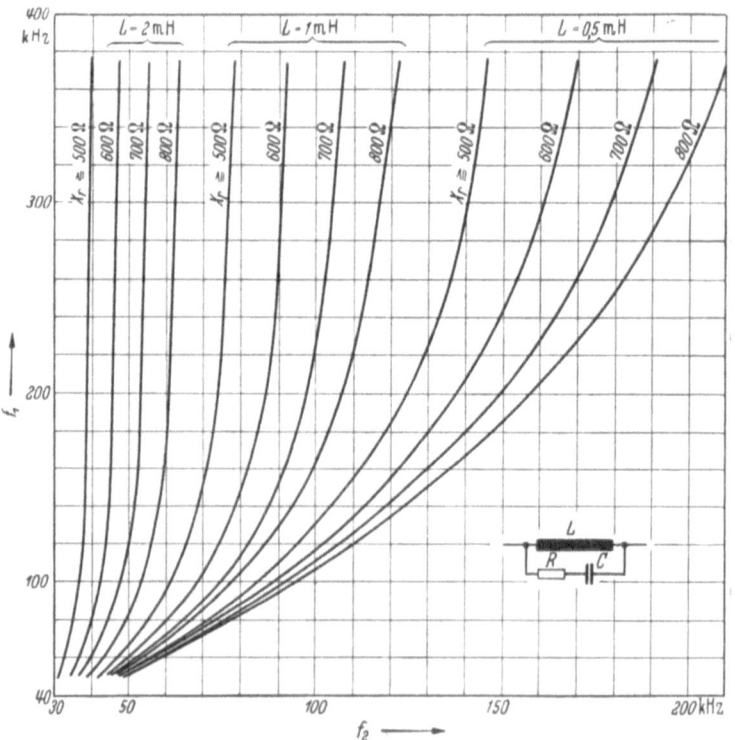

Fig. 78. Blocking range of broad-band traps as a function of inductance and minimum value of the ohmic component of the trap impedance, circuit arrangement 2

(f_1 = upper edge frequency, f_2 = lower edge frequency of blocking range)

This arrangement has a serious disadvantage. When an overcurrent flows through the coil as a result of a power line fault, the shunt resistor will have to carry the full voltage drop. Assuming a 2-mH coil and a 1000-Ω resistor, an overcurrent of, say, 5000 A will subject the resistor

to a temporary load of 10 kW. Using a 0.2 mH wave trap, this load will
be merely 100 W, all other conditions remaining unchanged. A circuit
arrangement avoiding this excessive load on the resistor is therefore
highly desirable. It will suffice to connect a capacitor ahead of the resistor. This capacitor acts as a reservoir for the entire voltage appearing
across the coil. To reduce its influence on wave trap tuning to a minimum,
it must be rated for a capacitance equalling or exceeding that of the
tuning capacitor.

Wave traps having an inductance of 1 mH or 2 mH can be tuned,
without any loss in blocking range, merely with the aid of the capacitor

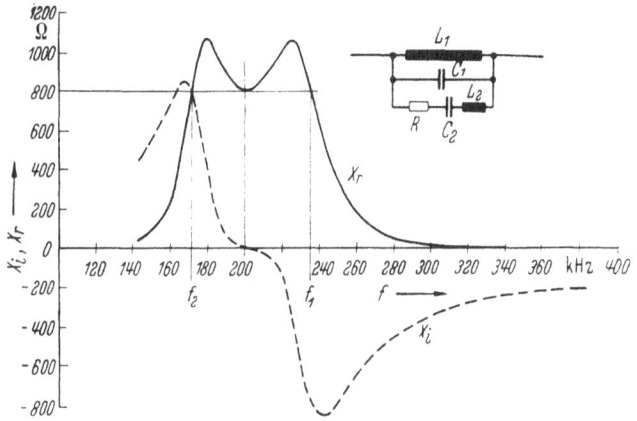

Fig. 79. Band tuning of a resonant trap.
$L_1 = 0.2$ mH, $L_2 = 2.0$ mH, $R = 800\ \Omega$, $f_2 = 172$ kHz, $f_0 = 201.5$ kHz, $f_1 = 236$ kHz

connected in series with the ohmic resistance (Fig. 78). A comparison
with the previous example shows that the blocking range obtained with
this circuit arrangement extends from $f_1 = 300$ kHz to $f_2 = 63$ kHz, i. e.,
that it is wider than the range available with the simple parallel arrangement (Fig. 77). At frequencies below 50 kHz the parallel arrangement
will yield better results for 2-mH wave traps.

Where more than two individual frequency bands are to be trapped
with the required ohmic resistance, the coils must not necessarily be
rated for an inductance of 1 mH or 2 mH. An alternative solution is
here broad-band tuning (Fig. 79). The parallel-resonant circuit L_1, C_1
and the series-resonant circuit L_2, C_2 are tuned to the same frequency f_0.
At the frequency f_0, the resistance of circuit L_2, C_2 is zero.

Thus, the trap impedance at this frequency is determined by the
value of R. The same ohmic resistance of the trap is obtained with
$f_1 > f_0 > f_2$, where $f_0 = |\ \overline{f_1 f_2}$. When optimizing L_2 so that $R^2 = \omega_0^2 \cdot L_1 L_2$

then the amount of the imaginary component X_i will equal the ohmic resistance R. A comparison between this arrangement and the simple circuit where the coil, the capacitor, and the resistor are connected in parallel, yields the following results:

Circuit arrangement	Fig. 77	Fig. 79
L mH	0.2	0.2
R Ω	1200	600
X_r Ω	600	600
f_2 kHz	160	160
f_1 kHz	193	235

9.3 Iron-Coated Power Conductors for Network Decoupling

Providing power conductors with an iron covering produces the effect of a crosstalk trap. As is generally known, alternating currents tend to concentrate on the surface of the conductors with rising frequencies (skin effect). If the top layer consists of a material having a high permeability, such as iron, and a conductivity below that of the inner conductor, a higher attenuation of the carrier currents will result. The cross-talk trapping effect of an iron covering of a power conductor may be explained by the increase of the ohmic resistance of the conductor which results from the skin effect.

Calculations yield a HF resistance of approximately 3.2Ω/mile at 100 kHz for a copper conductor having a diameter of 0.6 inches, whereas an iron conductor of the same diameter has a resistance of 160Ω/mile. The thickness of a layer, whose DC resistance equals the HF resistance of the solid conductor (equivalent conducting layer) amounts, in our example, to 0.01 inches for the copper conductor and to 0.001 inches for the iron conductor. At a thickness of 0.05 inches, which would have to be chosen for pure manufacturing reasons, conditions may be expected which approximate those obtained with a solid iron conductor.

Taking $Z = 700 \Omega$, the attenuation of a line $a = 8.7 \dfrac{R}{2Z}$ may be expressed as

$$a = 8.7 \frac{2.160}{2.900} \frac{\text{db}}{\text{miles}} = 2 \, \text{db/mile}$$

To achieve network decoupling with an attenuation of approximately 50 db, the copper conductor in question would have to be covered with iron over a length of 25 miles. At higher frequencies, a larger diameter of the conductor, and a smaller characteristic impedance, the length requiring an iron covering will decrease. The minimum length, however, will range between 1 and 6 miles, so that this network decoupling method stands little chance of practical application.

9.4 Decoupling by Four-Terminal Networks

A network decoupling circuit may be designed in the form of a regular four-terminal trap of the type generally used in filter engineering. The only difference is that the circuit components must be rated for use on power lines (Fig. 80). This arrangement permits two networks to be decoupled at their transition points so that the carrier currents are effectively blocked. It consists of a lowpass filter between each conductor and ground which allows currents of lower frequencies to pass practically without attenuation, while the attenuation rises sharply in the blocking range above the cut-off frequency.

If, taking the transmission range of the carrier terminals into consideration, an attenuation of 50 db were required over the entire frequency range of, say, 35 kHz to 375 kHz, multi-section lowpass filters with large

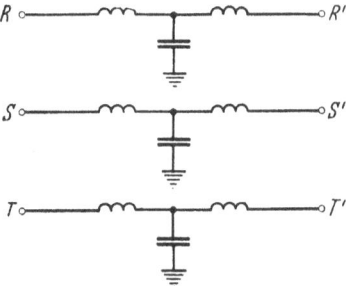

Fig. 80. Arrangement of a four-terminal circuit for network decoupling purposes.
R, S, T line side; R', S', T' station side

inductances and capacitances would have to be provided. Aside from the cost factor, the high values of L and C would change the parameters of the power line to the point where difficulties in power network protection would have to be expected. Sufficiently low values of L and C, and a corresponding cost reduction, can be achieved only by a suitable modification of the basic circuit. The latter will have to be designed so that the range to be trapped covers not the entire available frequency range but only the region beginning approximately at 100 kHz.

Moreover, the crosstalk attenuation need not necessarily amount to 50 db for all frequencies, because a value of about 35 db will also permit the same frequencies to be re-used at a distance equaling one line section.

Decoupling circuits of this type are not frequently used because of the comparatively high cost of the power capacitors and broad-band traps required for this purpose. Since such a project will benefit several utilities, a coordination of planning work and a cost sharing scheme will be required. In the case of a coupling line between the networks of two different companies a simplification can be achieved by distributing the decoupling circuit to the two ends of the line [15].

A more economical solution is obtained if the decoupling problem is tackled in connection with the measuring problem and the use of capacitive voltage transformers. This arrangement takes the edge off the chief argument against the introduction of network decoupling circuits, which is the high cost of a circuit serving no other but decoupling purposes. Under certain circumstances it might become necessary to rate the power capacitors for capacitance values higher than would be required for the measuring task alone (Fig. 81).

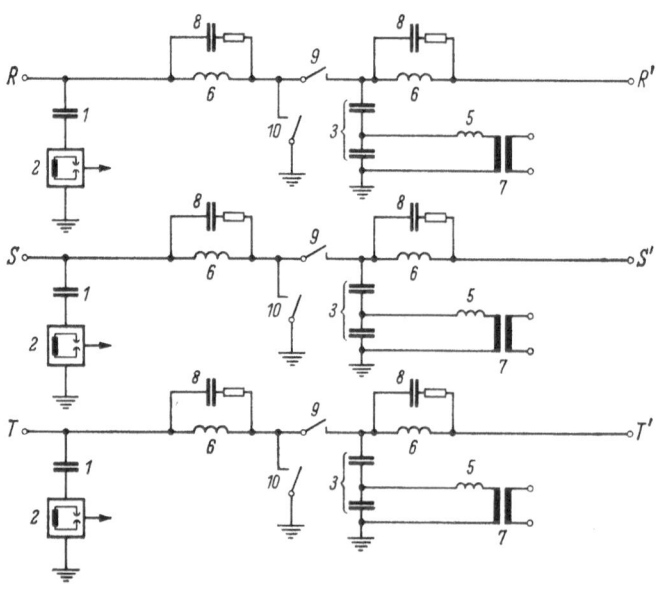

Fig. 81. Network decoupling circuit based on the common use of capacitive voltage transformers. *1* Coupling capacitor; *2* Line tuning unit; *3* Capacitive voltage divider; *4* Resistor; *5* HF choke coil; *6* Trap; *7* Auxiliary transformer; *8* Capacitor of four-pole circuit; *9* Disconnect switch; *10* Grounding switch; *R, S, T* power line side, *R', S', T'* station side

To avoid any quarrel about radio interference between the various communication systems, some countries, where hoarfrost is a factor which need not be considered, went so far as to bar the region below 150 kHz for carrier communication. This measure leads to a rapid crowding of the available frequency spectrum so that network decoupling by means of four-pole traps may very soon become a real necessity. It is here less of a problem because only the frequencies in the higher range have to be blocked.

9.5 Attenuation and Level

With the power transmitted at the beginning of a line denoted by P_1 and the power available at the end of the line by P_2, the "efficiency" η, as known in power engineering applications, can be defined by the expression

$$\eta = \frac{P_2}{P_1}.$$

The efficiency is always positive and less than one. In communication engineering on the other hand, the standard measure is the "attenuation":

$$a = \log \frac{P_1}{P_2}$$

The attenuation may be positive or negative (gain) and take on any numerical value. Attenuation and efficiency figures have no dimensions. While the efficiency is stated in percents or in the form of a decimal fraction, it is common practice to add the term "Bel" to attenuation values (after the inventor of the telephone: A.G. BELL).

To obtain numerical values which are more convenient for practical purposes, the tenth part of the value obtained from $\log P_1/P_2$ is taken for measurements. It is referred to as one "Decibel" (db)

$$a = 10 \log \frac{P_1}{P_2} (\text{db}).$$

The relationship between the efficiency and the attenuation is given by

$$\eta = \frac{P_2}{P_1} = 10^{\frac{a}{10}} \quad \text{or} \quad a = 10 \log \frac{1}{\eta} \, \text{db}.$$

In communication engineering, the efficiency is normally very small.

In some European countries another measure has been adopted for expressing the attenuation, namely

$$a = \frac{1}{2} \ln \frac{P_1}{P_2}$$

which is referred to as "Neper" (Np) (after the inventor of the natural logarithm: NAPIER). BRIGG's logarithm is here replaced by the natural logarithm.

From

$$10 \, \text{db} \cdot \log \frac{P_1}{P_2} = \frac{1}{2} \ln \frac{P_1}{P_2} \text{Np}$$

11 Podszeck, Carrier Communication, 4th Ed.

we obtain

$$1 = \mathrm{db} = \frac{\frac{1}{2} \ln \frac{P_1}{P_2}}{10 \log \frac{P_1}{P_2}} \, \mathrm{Np} = \frac{2.30259}{20} \, \mathrm{Np} = 0.115 \, \mathrm{Np}$$

and, inversely, 1 Np = 8.686 db.

Drawing on the terminology used for waterways, communication engineers speak of "levels" when they compare voltages or currents found at any given point of the circuit with the corresponding figures measured at another point of this circuit. The level denotes the logarithmic ratio of the voltages (or currents). Since the absolute magnitude of the starting level may be any value in dealing with ratios, the zero point on the level scale may be chosen arbitrarily. In accordance with an international agreement the level zero has been related to a voltage E_0 which, measured across an impedance $Z_0 = 600 \, \Omega$, will produce the power $P_0 = 1 \, \mathrm{mW}$. From

$$P_o = \frac{E_0^2}{Z_0} = 1 \, \mathrm{mW} \quad \text{we obtain} \quad E_o = \sqrt{P_o Z_o} = 0.775 \, \mathrm{V}$$

and from

$$P_o = I_0^2 Z_o = 1 \, \mathrm{mW} \quad \text{we obtain} \quad I_o = \sqrt{\frac{P_0}{Z_0}} = 129 \, \mathrm{mA} \, .$$

Level measurements are carried out with a "standard level oscillator" which, having an internal resistance of 600 Ω, supplies 1 mW across a 600-Ω load. Its e. m.f. amounts to $2 \cdot 0.775 \, \mathrm{V} = 1.55 \, \mathrm{V}$. When referring the value observed at the measuring point to the corresponding value at the beginning of the transmission path, the "relative level" will be obtained, whereas a comparison with the standardized reference value will yield the "absolute level". The values measured at any point along the circuit can be related to the zero point on the level scale with the aid of the equations

$$p = 10 \log \frac{P}{P_0} \, \mathrm{db} = 10 \log P \, \mathrm{db} \left(= \frac{1}{2} \ln P \, \mathrm{Np} \right)$$

for the power level (P in mW)

$$p_E = 20 \log \frac{E}{E_0} \, \mathrm{db} = 20 \log \frac{E}{0.775} \, \mathrm{db} \left(= \ln \frac{E}{0.775} \, \mathrm{Np} \right)$$

for the voltage level (E in V)

$$p_I = 20 \log \frac{I}{I_0} \, \mathrm{db} = 20 \log \frac{I}{1.29} \, \mathrm{db} \left(= \ln \frac{I}{1.29} \, \mathrm{Np} \right)$$

for the current level (I in mA).

An attenuation results in a decrease, an amplification in an increase of the level. A "level diagram" can be established by plotting the levels measured along the line, including coupling circuits, by-pass circuits. and amplifiers (Fig. 82). The magnitude of the level measured at any point of the line corresponds to the sum of attenuations and amplifications experienced up to this point.

The level diagram of a line with uniform impedance will be the same. both for the voltage and the power. To obtain a similar picture also in the case of a multi-section line having different impedance values at different

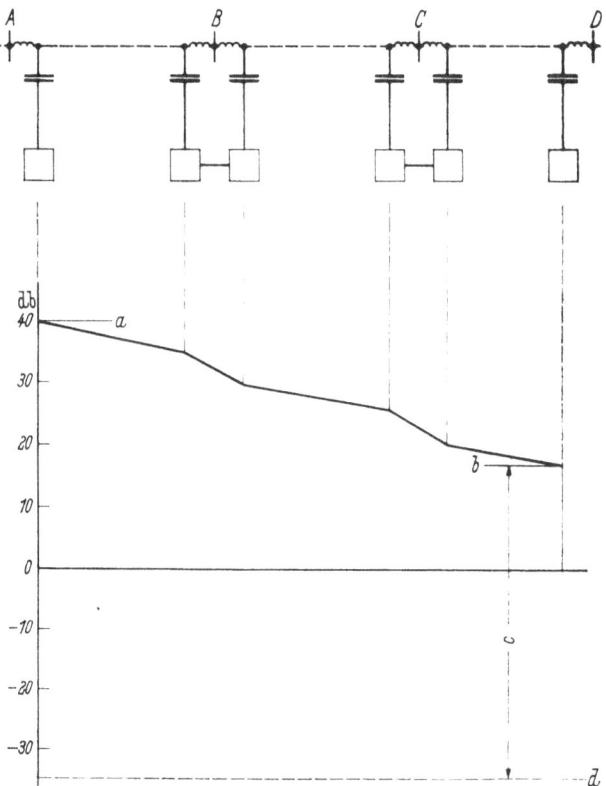

Fig. 82. Level diagram of a carrier frequency channel.
A Generator (transmitter); *B, C* By-pass circuits; *D* Load (receiver); *a* Send level; *b* Receive level; *c* Signal-to-noise ratio; *d* Noise level

points, it is normal practice to use only the power level for calculations. According to the established equations, the voltage level and the power level have numerically identical values on an open-wire line with an impedance of $Z = 600 \, \Omega$. At a given voltage $E = 15 \, \text{V}$, for instance (see page 177), the power level is given by

$$p = 10 \log P = 10 \log \left(\frac{E^2}{Z_0} \cdot 10^3 \right) = 10 \log \frac{225}{600} \cdot 1000 = +25.7 \, \mathrm{db}$$

$$\left[\frac{1}{2} \ln \left(\frac{225}{600} \cdot 10^3 \right) = +2.96 \, \mathrm{Np} \right]$$

and the voltage level by

$$p_E = 20 \log \frac{E}{E_0} \, \mathrm{db} = 20 \log \frac{15}{0.775} = +25.7 \, \mathrm{db}$$

$$\left(\ln \frac{15}{0.775} = +2.96 \, \mathrm{Np} \right).$$

If the voltage $E = 15 \, \mathrm{V}$ is measured on a cable having an impedance $Z = 140 \, \Omega$,
the power level is

$$p = 10 \log \left(\frac{15^2}{140} \cdot 10^3 \right) = 10 \log 1610 = +32.1 \, \mathrm{db}$$

$$\left(\frac{1}{2} \ln \frac{15^2}{140} \cdot 1000 = 3.69 \, \mathrm{Np} \right)$$

and the voltage level

$$p_E = 20 \log \frac{E}{E_0} \, \mathrm{db} = 20 \log \frac{15}{0.775} = +25.7 \, \mathrm{db}$$

$$\left(\ln \frac{15}{0.775} = +2.96 \, \mathrm{Np} \right).$$

When used in a general sense without any further definition, the term "level" refers to the power level. For comparing voltage and current values, a clear definition is required and the terms "voltage level" or "current level" must be given in full together with the value of the resistance across which the level has been measured.

The measurable quantities are mostly voltages and the voltage levels are therefore taken as level values. The power level p generally used in overall planning is computed from the voltage level p_E and the impedance Z as follows:

$$p = 10 \log \frac{P}{P_0}$$

$$P = \frac{E^2}{Z}; \ P_0 = \frac{E_0^2}{Z_0}$$

$$p = 20 \log \frac{E}{E_0} - 10 \log \frac{Z}{Z_0}$$

Since the expression $20 \log \dfrac{E}{E_0}$ represents the voltage level p_E, the power level differs from the voltage level merely by the correction factor

$$\varDelta_p = -10 \log \frac{Z}{600} \, \mathrm{db} = \left(-\frac{1}{2} \ln \frac{Z}{600} \, \mathrm{Np}\right)$$

which is given by the impedance deviation between Z and the standard value of $Z_0 = 600 \, \Omega$.

To determine the power level at any given value of Z from the voltage level, a diagram (Fig. 83) may be used which illustrates the correction factor $\varDelta_p$ as a function of the impedance deviation $Z/600$.

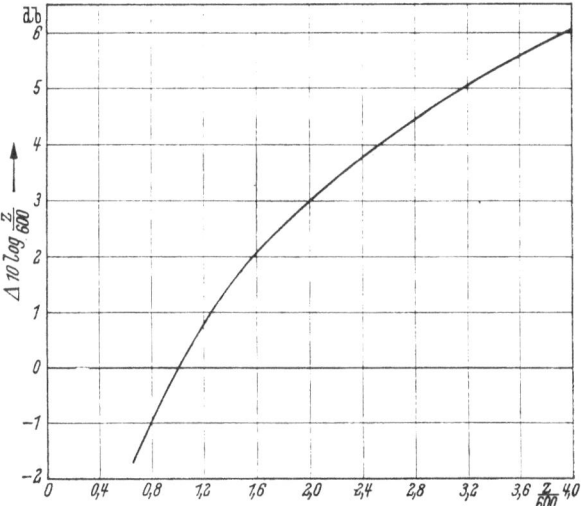

Fig. 83. Determination of power level from voltage level

Official regulations in many countries limit the permissible send power of power line carrier terminals to 10 W, measured at the input of the power line. This corresponds to a maximum permissible send level of

$$p_{\mathrm{max}} = 10 \log 10{,}000 = 40 \, \mathrm{db} \,.$$

With phase-to-ground coupling and $Z = Z_{ph\,gr} = 400 \, \Omega$:

$$E_{ph\,gr} = \sqrt{P_{\mathrm{max}} Z_{ph\,gr}} = \sqrt{4000} = 63.5 \, \mathrm{V}\,;\ p_E = 38 \, \mathrm{db}.$$

With phase-to-phase coupling and $Z = Z_{ph\,ph} = 700 \, \Omega$:

$$E_{ph\,ph} = \sqrt{P_{\mathrm{max}} Z_{ph\,ph}} = \sqrt{7000} = 84 \, \mathrm{V}\,;\ p_E = 41 \, \mathrm{db}.$$

9.6 Amplitude Modulation in Connection with Double-Sideband and Single-Sideband Transmission, Frequency Modulation

The high-frequency current generated in the transmitter serves as. a "carrier" for the intelligence to be transmitted and, for this purpose, is operated upon by voice currents or by current pulses. This process is. called "modulation". The carrier wave can be modulated by varying the amplitude, the frequency, or the phase angle. Accordingly, these pro-- cesses are termed "amplitude modulation", "frequency modulation", and "phase modulation". Additional modulation methods have been de-- veloped, which have been given the generic term pulse modulation.. They are, essentially, a high-speed signaling method used chiefly for speech transmissions.

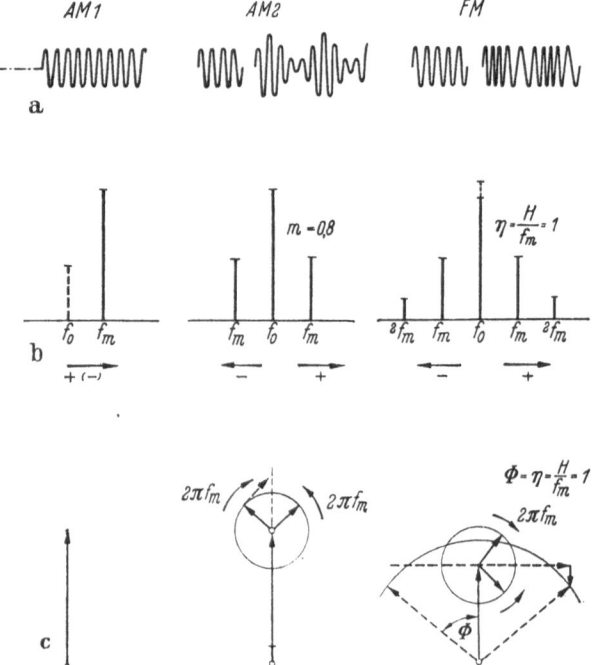

Fig. 84 a–c. Types of modulation.

a) HF wave plotted on a time basis; b) Amplitude pattern; c) Vector diagram; *AM 1* Amplitude-modulation with single-sideband transmission; *AM 2* Amplitude modulation with double-sideband transmission; *FM* Frequency modulation

The unmodulated carrier contains no information.. The actual message is contained in the HF frequencies produced by modulation at both sides of the carrier. Thus, a modulated carrier will take up a certain.

bandwidth. The various modulated carriers are separated by filter
circuits. The complete message transmission path is called "message
channel" or "transmission channel".

In theoretical deliberations a carrier is thought of as an auxiliary wave
required to shift the frequency band of signals from one frequency
position into another frequency position. Modulation and demodulation
are then visualized as a frequency conversion process.

Within the scope of this book amplitude modulation and frequency
modulation (Fig. 84) and the bandwidth occupied by a transmission
channel in either case are of interest.

In amplitude-modulated systems, the amplitude of the carrier frequency is varied in dependence of the strength of the modulating signal.
A single modulating frequency produces one sideband frequency each
at either side of the carrier, at a distance equaling the frequency of the
modulating signal. In the vector diagram, the two sideband vectors rotate
in opposite directions at the same angular velocity towards the carrier
vector whereby, due to their phase position, the resultant of both is
always either in the direction of the carrier, or away from the carrier.
This explains the amplitude variation of the resultant vector or the
resultant wave.

With double-sideband transmission, both the carrier and the two
sidebands are transmitted. Using the 300 Hz to 2400 Hz voice band for
modulation, a double-sideband channel will occupy a frequency band
of 5 kHz (Figs. 34 and 50b). With single-sideband transmission, on the
other hand, the frequency band assigned to a channel will be only half
as wide, i. e. 2.5 kHz (Fig. 34).

In frequency-modulated systems, the HF carrier is transmitted with
a constant amplitude. The carrier frequency is varied in proportion with
the instantaneous value of the modulating wave. The maximum deviation
from the carrier frequency, i. e. the frequency swing of the modulated
wave, corresponds to the amplitude of the modulating wave. This
variation in time of the carrier frequency results in an amplitude spectrum which, just as in amplitude modulation, exhibits two sidebands.
However, each modulating frequency f_m is here accompanied by a number
of sideband components spaced at a distance of f_m, $2 f_m$, $3 f_m$, $4 f_m$...
from the carrier frequency. Depending on the frequency swing and
the modulating frequency, the amplitudes of these sideband components
decrease more or less rapidly as their ordinal number increases. As distinct
from amplitude modulation, the amplitude of the carrier remains unchanged. The amplitudes are determined by the ratio of the frequency
swing S to the modulating frequency, that is, by the modulation index

$$\eta = \frac{S}{f_m}.$$

For a modulation index $\eta = 1$, taking a frequency swing of 2 kHz and a modulating frequency of 2 kHz as an example, the vector diagram in Fig. 84c shows the different phase angles of the sideband waves with amplitude modulation and frequency modulation. The two counter-rotating vectors of the first order sidebands, whose sum is in the direction of the carrier with amplitude modulation, result in a component perpendicular to the carrier in the case of frequency modulation. The higher order sidebands supplement the resulting carrier so that its peak rocks back and forth on an arc about its normal position. The amplitude therefore remains constant.

This swing of the carrier vector implies a change in its phase position. The phase deviation Φ, i. e. the maximum excursion of the carrier phase from its normal position, corresponds to the modulation index

$$\Phi = \eta = \frac{S}{f_m}.$$

This can be calculated from the relation between phase and frequency of the modulating signal. The instantaneous frequency corresponds to the differential quotient of the phase with respect to time. Frequency modulation and phase modulation are essentially identical and cannot be distinguished from each other if a single modulating frequency is used.

To assure a uniform frequency allocation scheme, the space occupied by a frequency-modulated transmission channel must not be wider than the space assigned to an amplitude-modulated double-sideband channel. In view of this limitation, the frequency swing must be made comparatively small (see page 62).

9.7 Send Power and Transmission Range of Carrier Terminals with the Different Transmission Methods[1]

The transmission range of a carrier circuit cannot be readily expressed in terms of miles. It is a function of the line attenuation and, consequently of the frequency used. Further, it depends on the configuration and on the condition of the line, and on the coupling method. Weather conditions are not normally an important factor, although hoarfrost may involve an appreciable deterioration. To span a certain distance, the signal-to-noise ratio at the receiver input must be made sufficiently large (Fig. 82). The send level of the carrier terminal and the transmission method employed are the determining factors for the transmission range in decibels, which can be obtained at a given signal-to-noise ratio. The corresponding distance in miles is therefore a function of the aforementioned parameters.

---

[1] According to E. ALSLEBEN.

The transmission range r, expressed as the permissible line loss a between the sending station and the receiving station, can be calculated from the send level p_{tr} and the allowable minimum receive level $p_{r\,\mathrm{min}}$.

$$r = a = p_{tr} - p_{r\,\mathrm{min}}. \tag{1}$$

The permissible minimum receive level depends on the noise level p_n and on the tolerable spacing Δp between the signal level and the noise level[1], both measured at the output of the receiver. Signals and noise level present at the receiver input may be evaluated differently by the receiver and their ratio may change between the input side (HF) and the output side (AF) of the receiver (during demodulation). Depending on the transmission method, the spacing between signal and noise may increase or decrease. As a measure of this change, a "gain factor" G or, on a logarithmic scale, a "gain of the transmission method" g may be used which, at a given transmitter power, results in an increase or in a decrease of the transmission range, depending on the transmission method employed.

$$G = \sqrt{\left(\frac{P_s}{P_n}\right)_{\mathrm{AF}} : \left(\frac{P_s}{P_n}\right)_{\mathrm{HF}}}; \quad g = 10 \log G = (\Delta p)_{\mathrm{AF}} - (\Delta p)_{\mathrm{HF}}. \tag{2}$$

Hence, the transmission range may be expressed as

$$r = p_{tr} - (p_n)_{\mathrm{HF}} - (\Delta p)_{\mathrm{AF}} + g, \tag{3}$$

or, if power ratios are used instead of levels,

$$r = 10 \left[\log\left(\frac{P_s}{P_n}\right)_{\mathrm{HF}} - \log\left(\frac{P_s}{P_n}\right)_{\mathrm{AF}} + \log\left(\frac{P_s}{P_n}\right)_{\mathrm{AF}} : \left(\frac{P_s}{P_n}\right)_{\mathrm{HF}} \right]. \tag{4}$$

Fixed standards, empirical values, and relationships have been established and compiled for the individual factors.

a) Transmitter Power

The send power or rated power P_{tr} of a transmitter is the carrier energy of the unmodulated transmitter in the case of frequency-modulated carrier terminals and in the case of amplitude-modulated double-sideband terminals. In single-sideband carrier terminals, however, it refers to the energy content of one sideband with the transmitter fully loaded with a single frequency. In an amplitude- or frequency-modulated signaling channel it is the output measurable at steady tone or at one of the two shift frequencies.

---

[1] The difference between noise voltage and "psophometric voltage" (according to the CCITT weighting curve, or as compared with a standard 800 Hz noise signal), referring only to voice channels, is here of little significance and need not be considered.

In frequency-modulated systems the signal strength remains constant and the send amplifier need be rated only for the rated power. In an amplitude-modulated transmitter, on the other hand, the send amplifier must be rated for a peak power higher than the nominal output.

For the single-sideband case, this peak power is given by the maximum value of the beat between the useful signal, transmitted at the nominal power, and the usually transmitted pilot tone or residual carrier. If the amplitude of the latter amounts to q per cent of the useful signal with full control, the peak power will be

$$P_{p,\,ss} = \left(1 + \frac{q}{100}\right)^2 P_{tr}\,. \tag{5}$$

Assuming a pilot tone amounting to 20 per cent of the maximum useful signal amplitude, the required peak power would be $1.44 \times P_{tr}$.

In double-sideband transmission, the amplifier must be rated for the beat between the carrier and the two sidebands. From the vector diagram of the modulation methods (Fig. 84c) one may obtain the voltage during the maximum beat and, consequently, the power

$$P_{p,\,ds} = (1 + m)^2 P_{tr} \tag{6}$$

and the energy content of one sideband

$$P_{sb,\,ds} = \frac{m^2}{2} P_{tr} \tag{7}$$

With a modulation factor $m = 100$ per cent, the amplifier would have to be rated for a peak power four times as high as the nominal power. The sum of the power components transmitted over the two sidebands, which alone contain the actual message, would then be only half the nominal power. In conventional-type carrier terminals, however, the modulation factor obtained at full load varies between 60 per cent and 80 per cent. At a modulation factor of 80 per cent, the peak power will amount to $3.24 \cdot P_s$. Compared to other types of modulation, a rather high output is required for double-sideband systems. In our by no means unfavorable example it amounts to approximately ten times the power effectively transmitted over the two sidebands.

In multi-channel signaling applications the peak power is

$$P_{p,\,ch} = n^2 P_{tr} \tag{8}$$

where n is the number of channels and P_{tr} the send power of the un-modulated channel carrier. The peak power is determined by the amplifier rating. The send power P_{tr} available for each channel decreases with the square of the number of channels. If the channel group is transmitted on a double-sideband instead of a single-sideband basis, the modulation factor is divided linearly with the number of channels (the sideband power on a square law basis). The channel power may be slightly increased where a larger number of channels must be transmitted and if it is assumed that the maximum beat will not occur in all channels simultaneously. It is equally irrelevant to assume that all channels will persistently be keyed in the same direction, a condition which, in amplitude-modulated systems, would produce the same result.

In multi-purpose equipment designed for the simultaneous transmission of speech and superimposed signaling channels, the total power may be apportioned so that, with increasing line attenuation, the permissible signal-to-noise ratios are obtained for speech and signaling channels simultaneously. Where 6 signaling channels of a bandwidth of 80 Hz (120 Hz channel spacing) are transmitted, the peak power will be assigned to speech and signaling channels at a ratio of approximately 1 : 1, so that one quarter of the peak power will be available for either application. Each superimposed signaling channel will then receive a power equalling $1/n^2 \cdot 1/4$ of the peak power, that is, 1/144 of the total available peak power in the 6 channel case.

In many countries the output power of a transmission terminal has been fixed at 10 W. This applies, in agreement with the previously given definition of the term "rated power", to the sideband power obtained under full load in single-sideband equipment, and to the unmodulated carrier power in the case of double-sideband equipment and frequency-modulated equipment. In addition to this, regulations provide for a permissible peak power of 40 W. In double-sideband equipment with a carrier independent of the modulation factor and a carrier power of 10 W this peak power is obtained, according to equation (5), if the depth

Number of channels n		1	2	4	6	12	18	
Channel power	P_{ch}	10	5	2.5	1.1	0.28	0.12	watts
Total available power	$n\,P_{ch}$	10	10	10	6.6	3.4	2.2	watts
Peak power	$n^2\,P_{ch}$	10	20	40	40	40	40	watts

Fig. 85. Permissible channel power in multi-channel equipment for the transmission of pulsed information.

of modulation is made 100 per cent. With multi-channel equipment for signaling applications this 10 W limit applies to the sum of the power portions assigned to the individual channels within a 2.5 kHz band. It determines the permissible channel power for 1 to 4 channels. Where more than 4 channels are transmitted, the 40-W limit will be of interest. From this can be computed the permissible individual-channel energy as a function of the number of the channels to be transmitted (Fig. 85).

Another important point in determining the transmission range of carrier telephone terminals is the speech level assigned to the full-load condition. In view of the high noise level on power lines it is normal practice to limit the speech level to 1 mW at the relative level of zero and to use this power for full control. In postal systems, however, this limit has been fixed at + 7 db. Adopting this regulation for power line carrier communication systems would mean a decrease of 7 db in the possible transmission range.

b) Noise Level and Signal-to-Noise Ratio

The transmission range of carrier terminals is limited by the permanent random noise, corona noise in the first place, which is associated with power current transmission. Selective noise introduced by interfering radio transmitters can be avoided by selecting suitable frequencies and by previous frequency allocation arrangements. Impulse noise caused by the operation of power switches, by atmospheric discharges, and by power line faults, can be adequately suppressed through limitation in the coupling facilities and in the receivers as well as proper selection of time constants and similar measures. In protective carrier relaying, where the signal transmission times are extremely short and reliability requirements very high, this impulse-type noise decides on the extent to which the transmission range, available after taking into account the permanent noise, can actually be exploited.

The empirical values found for corona noise in a 2.5 kHz frequency band (see page 59) can be correlated to other frequency bands and used as a starting point for calculating the transmission range. A signal-to-noise ratio of 26 db is normally deemed permissible as a lower limit both for speech and the 80 Hz signaling channels. It will be advisable to add about 9 db as a level reserve to allow for abnormal line conditions.

c) Gain Factor of Transmission Method

By way of approximation, corona noise may be visualized as having a spectrum with frequency-independent amplitude e_n within the HF transmission band. The value e_n is to be viewed as the noise voltage referred to the unit of the frequency-band, from which the noise voltage E_n and the noise power P_n of the total band can be computed across resistor R as

$$E_n = e_n \sqrt{\Delta f}; \quad P_n = \frac{e_n^2 \Delta f}{R}. \tag{9}$$

For single-sideband applications, all HF voltages within the working band, that is, useful signal voltages and unwanted noise voltages, are converted alike to the voice-frequency position. The signal-to-noise voltage ratio and the signal-to-noise power ratio remain unaffected.

Hence,

$$\left(\frac{P_s}{P_n}\right)_{\mathrm{AF}} = \left(\frac{P_s}{P_n}\right)_{\mathrm{HF}} \tag{10}$$

and, consequently,

$$G_{ss} = 1; \quad g_{ss} = 0 \,\mathrm{db}. \tag{11}$$

This gain factor is a handy reference for any comparison between the different transmission methods.

In double-sideband operation it is convenient, in accordance with the definition of the send power, to regard the carrier power P_{tr} or the carrier voltage E_{tr} as the useful signal power or useful signal voltage available across the input of the receiver. However, the useful signal voltage which becomes effective at the output of the receiver after the demodulation process is merely the result of the in-phase addition of the two sideband vectors (Fig. 84c). It is $E_s = m_0 E_{tr}$, while the effective noise voltage is given by $|\overline{2}E_n = |\overline{2}e_n|\Delta f$ as the resultant noise voltage of both sidebands. The frequency $f_{\max}$ may be inserted in the place of the bandwidth Δf, i. e. the highest transmitted frequency of a band ranging from zero to $f_{\max}$. The signal-to-noise voltage ratio then becomes

$$\left(\frac{E_s}{E_n}\right)_{\mathrm{AF}} = \frac{m_0}{\sqrt{2}}\left(\frac{P_s}{P_n}\right)_{\mathrm{HF}} \tag{12}$$

and the corresponding power ratio

$$\left(\frac{P_s}{P_n}\right)_{\mathrm{AF}} = \frac{m_0^2}{2}\left(\frac{P_s}{P_n}\right)_{\mathrm{HB}}. \tag{13}$$

Thus, the gain factor and the gain are:

$$G_{ds} = \frac{m_0}{\sqrt{2}}; \quad g_{ds} = 20 \log \frac{m_0}{\sqrt{2}}. \tag{14}$$

At a maximum modulation factor of $m_0 = 0.8$, the values of $G_{ds} = 0.57$ and $g_{ds} = -5$ db will be obtained. These figures show the disadvantage of double-sideband operation as compared to the single-sideband technique.

In frequency-modulated systems the ratio of the signal frequency swing S_0 to the noise frequency swing S_n corresponds to the ratio of the useful signal voltage to the noise signal voltage at the output of the equipment. This ratio can be derived from a vector diagram (Fig. 86) in the well-known manner. The carrier voltage vector is here supplemented by a rotating noise voltage vector (the noise voltage e_n is referred to a bandwidth of 1 cps), deviating from the carrier by frequency f_n so as to form a resultant vector for the carrier and the noise voltage. This vector rocks back and forth through an angle corresponding to the noise phase angle φ_n. When e_n is smaller than E_s, a fair approximation is given by the formula

$$\varphi_n = \frac{e_n}{E_s}. \tag{15}$$

The noise frequency swing can be computed from the noise phase swing as

$$S_n = f_n \varphi_n = f_n \frac{e_n}{E_s} \tag{16}$$

It follows that the noise frequency swing is proportional to the spacing between the noise signal frequency and the useful signal frequency. This spacing equals the frequency f_n of the noise voltage present after demodulation.

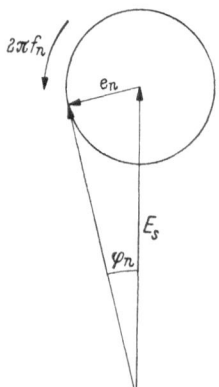

Fig. 86. Addition of noise and carrier voltages with frequency modulation

Since the noise voltage is made up of components from both sidebands, its magnitude increases by the factor $\sqrt{2}$, just as in the double-sideband case, and the signal-to-noise voltage ratio may be expressed as

$$\left(\frac{E_s}{e_n}\right)_{\mathrm{AF}} = \frac{S_0}{\sqrt{2}\,S_n} = \frac{S_0}{\sqrt{2}\,f_n} \cdot \left(\frac{E_s}{e_n}\right)_{\mathrm{HF}}. \tag{17}$$

Like the noise frequency swing, the noise voltage at the output of the receiver varies in direct proportion with the frequency. The same applies to the signal-to-noise ratio, i. e. to the reciprocal of the gain factor $(1/G)$ which, according to equation (17) is given by $\frac{f_n}{S_0}\sqrt{2}$. Assuming $f_{\max} = S_0$, its value will be $\sqrt{2}$, a value which, according to equation (14), would be obtained in the entire voice frequency range for double-sideband operation with $m = 100$ per cent. The reciprocals of the gain factors of the three transmission methods can best be compared with the aid of a diagram (Fig. 87). In frequency modulation, the gain factor is a function of the frequency and a direct comparison is therefore possible only for sufficiently narrow frequency bands, or for a median value obtained from the power in the entire voice frequency band by integration from $f = 0$ to $f = f_{max}$. Corresponding to the linear voltage rise, the power rises in a

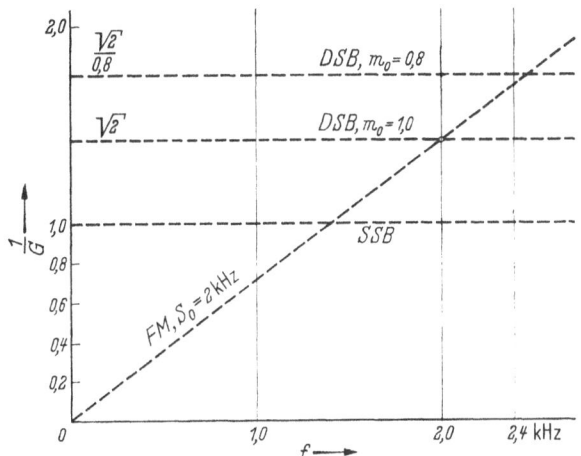

Fig. 87. Reciprocal of gain factor as a function of frequency with amplitude modulation and frequency modulation

square fashion with the frequency, and the ratio of signal power to noise power is obtained by integration as

$$\left(\frac{P_s}{P_n}\right)_{AF} = \left(\frac{S_0}{\sqrt{2}\, f_{max}}\right)^2 \cdot \frac{E_s^2}{\frac{1}{3}\, e_n^2 f_{max}} = \frac{3}{2}\left(\frac{S_0}{f_{max}}\right)^2 \cdot \left(\frac{P_s}{P_n}\right)_{HF}. \tag{18}$$

From this can be derived the mean gain factor for frequency modulation:

$$\bar{G}_{fm} = \frac{S_0}{f_{max}}\sqrt{\frac{3}{2}} \tag{19}$$

and

$$\bar{g}_{fm} = 20\log\frac{S_0}{f_{max}}\sqrt{\frac{3}{2}} \tag{20}$$

In carrier communication over power lines narrow frequency bands are used also for frequency-modulated channels. With a frequency swing of 2 kHz and a transmitted AF band of up to 2.4 kHz ,the gain factor will still approximate 1. This corresponds to 0 decibels. It may therefore be said that, assuming equal send power, there is no difference between frequency modulation and single-sideband transmission as far as the signal-to-noise ratio is concerned.

d) Practical Application

The gain factors have been computed on the basis of random noise, with the noise power equally distributed over the frequency band, and from the resultant noise power P_n in a HF band Δf, whose width equals that of the voice-frequency channel. When calculating the transmission range according to equation (2), it will therefore be necessary to insert always a noise power P_n or a corresponding noise level p_n, correlated to a HF band of this width.

Taking the above considerations as a basis for compiling the transmission range afforded by the different carrier terminals, a survey (Fig. 88) will be obtained as a planning basis for carrier transmission systems. The noise levels actually found in practice may depart considerably from these calculated values (see page 59). In marginal cases it will therefore be advisable to determine the actual noise levels by measurement and to use the measured values for calculating the transmission range.

	Frequency modulation	Telephone terminals		Amplitude modulation Multi-channel signaling terminals				
		DSB	SSB					
Number of channels				2	4	6	12	18
Send level p_{tr}	+40	+40	+40	+37	34	30	24	17
Gain g	0	− 5	0	−	−	−	−	−
SNR Δp	26	26	26	26	26	26	26	26
$p_{tr}+g-\Delta p$	+14	+ 9	+14	+11	+ 8	+ 4	− 2	− 9
p_n (220 kV)	−21	−21	−21	−35	−35	−35	−35	−35
Transm. range r_{220}	35	30	35	46	43	39	33	26
p_n (110 kV)	−38	−38	−38	−52	−52	−52	−52	−52
Transm. range r_{110}	52	47	52	63	60	56	50	43

Fig. 88. Transmission range of power line carrier terminals

Send power	10 W
Peak power	40 W for $n \geq 4$
Modulation factor under full load double-sideband operation	80 per cent
Frequency swing under full load frequency modulation	± 2 kHz
Noise level of 220 kV line	− 17 db/5 kHz
Noise level of 110 kV line	− 35 db/5 kHz

9.8 Table of the 4-Digit Mantissae of Briggs' Logarithms 100 — 549

Number	0	1	2	3	4	5	6	7	8	9	D
10	0000	0043	0086	0128	0170	0212	0253	0294	0334	0374	40
11	0414	0453	0492	0531	0569	0607	0645	0682	0719	0755	37
12	0792	0828	0864	0899	0934	0969	1004	1038	1072	1106	33
13	1139	1173	1206	1239	1271	1303	1335	1367	1399	1430	31
14	1461	1492	1523	1553	1584	1614	1644	1673	1703	1732	29
15	1761	1790	1818	1847	1875	1903	1931	1959	1987	2014	27
16	2041	2068	2095	2122	2148	2175	2201	2227	2253	2279	25
17	2304	2330	2355	2380	2405	2430	2455	2480	2504	2529	24
18	2553	2577	2601	2625	2648	2672	2695	2718	2742	2765	23
19	2788	2810	2833	2856	2878	2900	2923	2945	2967	2989	21
20	3010	3032	3054	3075	3096	3118	3139	3160	3181	3201	21
21	3222	3243	3263	3284	3304	3324	3345	3365	3385	3404	20
22	3424	3444	3464	3483	3502	3522	3541	3560	3579	3598	19
23	3617	3636	3655	3674	3692	3711	3729	3747	3766	3784	18
24	3802	3820	3838	3856	3874	3892	3909	3927	3945	3962	17
25	3979	3997	4014	4031	4048	4065	4082	4099	4116	4133	17
26	4150	4166	4183	4200	4216	4232	4249	4265	4281	4298	16
27	4314	4330	4346	4362	4378	4393	4409	4425	4440	4456	16
28	4472	4487	4502	4518	4533	4548	4564	4579	4594	4609	15
29	4624	4639	4654	4669	4683	4698	4713	4728	4742	4757	14
30	4771	4786	4800	4814	4829	4843	4857	4871	4886	4900	14
31	4914	4928	4942	4955	4969	4983	4997	5011	5024	5038	13
32	5051	5065	5079	5092	5105	5119	5132	5145	5159	5172	13
33	5185	5198	5211	5224	5237	5250	5263	5276	5289	5302	13
34	5315	5328	5340	5353	5366	5378	5391	5403	5416	5428	13
35	5441	5453	5465	5478	5490	5502	5514	5527	5539	5551	12
36	5563	5575	5587	5599	5611	5623	5635	5647	5658	5670	12
37	5682	5694	5705	5717	5729	5740	5752	5763	5775	5786	12
38	5798	5809	5821	5832	5843	5855	5866	5877	5888	5899	12
39	5911	5922	5933	5944	5955	5966	5977	5988	5999	6010	11
40	6021	6031	6042	6053	6064	6075	6085	6096	6107	6117	11
41	6128	6138	6149	6160	6170	6180	6191	6201	6212	6222	10
42	6232	6243	6253	6263	6274	6284	6294	6304	6314	6325	10
43	6335	6345	6355	6365	6375	6385	6395	6405	6415	6425	10
44	6435	6444	6454	6464	6474	6484	6493	6503	6513	6522	10
45	6532	6542	6551	6561	6571	6580	6590	6599	6609	6618	10
46	6628	6637	6646	6656	6665	6675	6684	6693	6702	6712	9
47	6721	6730	6739	6749	6758	6767	6776	6785	6794	6803	9
48	6812	6821	6830	6839	6848	6857	6866	6875	6884	6893	9
49	6902	6911	6920	6928	6937	6946	6955	6964	6972	6981	9
50	6990	6998	7007	7016	7024	7033	7042	7050	7059	7067	9
51	7076	7084	7093	7101	7110	7118	7126	7135	7143	7152	8
52	7160	7168	7177	7185	7193	7202	7210	7218	7226	7235	8
53	7243	7251	7259	7267	7275	7284	7292	7300	7308	7316	8
54	7324	7332	7340	7348	7356	7364	7372	7380	7388	7396	8

Column D shows the difference between the last log. in a line and the first log. of the next following line.

9. Appendix

Table of the 4-Digit Mantissae of Briggs' Logarithms 550—999

Number	0	1	2	3	4	5	6	7	8	9	D
55	7404	7412	7419	7427	7435	7443	7451	7459	7466	7474	8
56	7482	7490	7497	7505	7513	7520	7528	7536	7543	7551	8
57	7559	7566	7574	7582	7589	7597	7604	7612	7619	7627	7
58	7634	7642	7649	7657	7664	7672	7679	7686	7694	7701	8
59	7709	7716	7723	7731	7738	7745	7752	7760	7767	7774	8
60	7782	7789	7796	7803	7810	7818	7825	7832	7839	7846	7
61	7853	7860	7868	7875	7882	7889	7896	7903	7910	7917	7
62	7924	7931	7938	7945	7952	7959	7966	7973	7980	7987	6
63	7993	8000	8007	8014	8021	8028	8035	8041	8048	8055	7
64	8062	8069	8075	8082	8089	8096	8102	8109	8116	8122	7
65	8129	8136	8142	8149	8156	8162	8169	8176	8182	8189	6
66	8195	8202	8209	8215	8222	8228	8235	8241	8248	8254	7
67	8261	8267	8274	8280	8287	8293	8299	8306	8312	8319	6
68	8325	8331	8338	8344	8351	8357	8363	8370	8376	8382	6
69	8388	8395	8401	8407	8414	8420	8426	8432	8439	8445	6
70	8451	8457	8463	8470	8476	8482	8488	8494	8500	8506	7
71	8513	8519	8525	8531	8537	8543	8549	8555	8561	8567	6
72	8573	8579	8585	8591	8597	8603	8609	8615	8621	8627	6
73	8633	8639	8645	8651	8657	8663	8669	8675	8681	8686	6
74	8692	8698	8704	8710	8716	8722	8727	8733	8739	8745	6
75	8751	8756	8762	8768	8774	8779	8785	8791	8797	8802	6
76	8808	8814	8820	8825	8831	8837	8842	8848	8854	8859	6
77	8865	8871	8876	8882	8887	8893	8899	8904	8910	8915	6
78	8921	8927	8932	8938	8943	8949	8954	8960	8965	8971	5
79	8976	8982	8987	8993	8998	9004	9009	9015	9020	9025	6
80	9031	9036	9042	9047	9053	9058	9063	9069	9074	9079	6
81	9085	9090	9096	9101	9106	9112	9117	9122	9128	9133	5
82	9138	9143	9149	9154	9159	9165	9170	9175	9180	9186	5
83	9191	9196	9201	9206	9212	9217	9222	9227	9232	9238	5
84	9243	9248	9253	9258	9263	9269	9274	9279	9284	9289	5
85	9294	9299	9304	9309	9315	9320	9325	9330	9335	9340	5
86	9345	9350	9355	9360	9365	9370	9375	9380	9385	9390	5
87	9395	9400	9405	9410	9415	9420	9425	9430	9435	9440	5
88	9445	9450	9455	9460	9465	9469	9474	9479	9484	9489	5
89	9494	9499	9504	9509	9513	9518	9523	9528	9533	9538	4
90	9542	9547	9552	9557	9562	9566	9571	9576	9581	9586	4
91	9590	9595	9600	9605	9609	9614	9619	9624	9628	9633	5
92	9638	9643	9647	9652	9657	9661	9666	9671	9675	9680	5
93	9685	9689	9694	9699	9703	9708	9713	9717	9722	9727	4
94	9731	9736	9741	9745	9750	9754	9759	9763	9768	9773	4
95	9777	9782	9786	9791	9795	9800	9805	9809	9814	9818	5
96	9823	9827	9832	9836	9841	9845	9850	9854	9859	9863	5
97	9868	9872	9877	9881	9886	9890	9894	9899	9903	9908	4
98	9912	9917	9921	9926	9930	9934	9939	9943	9948	9952	4
99	9956	9961	9965	9969	9974	9978	9983	9987	9991	9996	4

Column D shows the difference between the last log. in a line and the first log. of the next following line.

Bibliography

A. Books

1. CEI: Specification pour condensateurs de réseau, 1ère–3ième Partie, Genève (Suisse): Bureau Central de la CEI, 1954, 1955, 1957.
2. CHEVALLIER, A.: Télétransmission par ondes porteuses dans les réseaux de transport d'énergie à haute tension, Paris: Dunod 1946.
3. DRESSLER, G.: Hochfrequenz-Nachrichtentechnik für Elektrizitätswerke, Berlin: Springer 1941.
4. HENNING, W.: Die Fernwirktechnik im Dienste der Elektrizitätsversorgung, 3. Aufl., München/Wien: Oldenbourg 1963.
5. JOHN, S., BERGMANN, G.: Die Fernmessung Bd. III, Karlsruhe: G. Braun 1963.
6. ZUR MEGEDE, W.: Fortleitung elektrischer Energie längs Leitungen in Starkstrom- und Fernmeldetechnik, Berlin/Göttingen/Heidelberg: Springer 1950.
7. NEUGEBAUER, H.: Selektivschutz, 2. Aufl., Berlin/Göttingen/Heidelberg: Springer 1958.
8. SCHÖNHAMMER, K., VOSS, H. H.: Fernschreibübertragungstechnik, München/Wien: Oldenbourg 1966.
9. SWOBODA, G.: Die Planung von Fernwirkanlagen, München/Wien: Oldenbourg 1967.
10. UCPTE: Der Schutz grenzüberschreitender Kuppelleitungen unter Verwendung von TFH-Signalverbindungen für den Leitungsschutz, Laufenburg (Schweiz), 1959/60, Jahresbericht.

C. Papers Published in Engineering Journals, Reports

11. AIEE: Guide to application and treatment of channels for power-line carrier. AIEE Transactions 73 (1954) Pt III-A, pp. 417–436.
12. ALSLEBEN, E.: Betriebsdämpfung von Hochspannungsleitungen im Trägerfrequenzbereich, ETZ A 80 (1959) No. 8, pp. 245–251.
13. ALSLEBEN, E.: Valeurs des grandeurs caractéristiques des réseaux à haute tension à prendre en considération pour l'établissement d'un projet d'installation à courants porteurs et données pour la mesure de ces grandeurs. CIGRE 1962, Rapport 319.
14. ALSLEBEN, E., SCHUMM, E.: Schutzsignalübertragung über Hochspannungsleitungen, ETZ B 19 (1967) No. 4, pp. 89–95.
15. ALSLEBEN, E., BERGMANN, G.: Entkopplung von Hochspannungsnetzen für die Trägerfrequenzübertragung (TFH). Österr. Zeitschrift für Elektrizitätswirtschaft, Vol. 23 (1970) No. 4, pp. 127 bis 131.
16. BARSTOW, J. M.: Carrier telephones for farms. Bell Labor. Rec. 25 (1947) No. 10, pp. 363–366.

17. CHEVALLIER, A., HOLLEVILLE, M., BARRAULT, F.: Perturbations produites par les ondes parasites engendrées par les manoeuvres des sectionneurs. CIGRE, Session 1950, Rapport 337.

18. CHEVALLIER, A.: Prédétermination des conditions de propagation d'une onde à haute fréquence, se propageant le long d'une ligne triphasée symétrique à haute tension, lorsque le générateur de cette onde attaque la ligne entre un conducteur de phase et la terre. Rev. Gén. Electr. 60 (1951) pp. 164–172.

19. ELMUND, H., ENGSTRÖM, S., HOLLNER, A.: Recherches expérimentales sur les interférences transitoires provenant des défauts des lignes d'énergie sur l'equipement à courant porteur, du point de vue des relais. CIGRE, Session 1952, Rapport 303.

20. FLEISCHER, H., GRAFF, G.: Abschätzung der höchstzulässigen Fremdspannungen für Trägerfrequenzverbindungen auf Hochspannungsleitungen und Vorschläge für Mindestempfangspegel der hochfrequenten Signale. Elektrizitätswirtsch. 58 (1959) No. 22, pp. 775–781.

21. FLEISCHHAUER, W., PODSZECK, H. K., VOGL, W.: Vielfach-Trägerfrequenz-Nachrichtenübertragung über Hochspannungs-Bündelleiter. Siemens Zeitschr. No. 8 (1963), pp. 589–596.

22. GROSSKOPF, J.: Die Beeinflussung des trägerfrequenten Sprechverkehrs auf Hochspannungsleitungen durch Funksender. FTZ 7 (1954) No. 11, pp. 623–636.

23. HABANN, E.: Die HF-Telephonie und ihre Drosseleinrichtungen. Elektrizitätswirtsch. 27 (1928) Okt. 1/468, pp. 499–505.

24. HOCHRAINER, H.: Fortschritte auf dem Gebiete der Rundsteuerung von Elektrizitätsversorgungsnetzen. Elektrotechnik und Maschinenbau 86 (1969) No. 2, pp. 44–65.

25. JAUDET, R.: Origine, nature et ordre de grandeur des oscillations à haute fréquence produites par les manoeuvres de sectionneurs sur les réseaux de transport d'énergie à haute tension. Bull. Soc. Franc. Electr. 8 série, tome I, No. 6, (Juin 1960) pp. 381–398.

26. KALCKHOFF, G.: Über die Hochfrequenzverluste in einer Hochspannungsschaltstation beim Trägerfrequenzfernsprechen auf Hochspannungsleitungen. Frequenz 7 (1953), No. 1, pp. 1–8.

27. KUHN-KUNIEWSKI, H.: Influence des transformateurs de puissance sur la distribution de courants porteurs dans les lignes de transport d'énergie à haute tension. CIGRE, Session 1954, Rapport 315.

28. LUTSCH, A.: Beitrag zur Frage der Anwendbarkeit des Einseitenbandverfahrens für frequenzmodulierte Schwingungen. FTZ 2 (1949), No. 11, pp. 347–351.

29. MATTHIES, H.: Trägerfrequenz-Übertragung auf isolierten Erdseilen von Hochspannungsleitungen in den USA. ETZ Ed. B (1967), No. 26, pp. 737–741.

30. MIKKELSEN, M. A.: Affaiblissement de la transmission par courant porteurs sur lignes d'énergie, specialement pendant la formation de verglas et de givre sur les conducteurs. CIGRE, Session 1950, Rapport 323.

31. MOEBES, R.: EW-Fernmeldeanlagen als Ursache von Rundfunkstörungen. Telegr. Prax. 18 (1938) No. 10, pp. 153/154.

32. MOEBES, R.: Gegenseitige Beeinflussung von drahtgebundenen und drahtlosen Funkdiensten. ETZ 63 Ed. A (1942) No. 9/10, pp. 113–116.

33. PELISSIER, R.: Propagation des ondes éléctromagnétiques guidées par une ligne multifilaire. Rev. Gen. Electr. 78 (1969) pp. 337–352, pp. 491–505.

34. PODSZECK, H. K.: Die Bandbreite von Trägerfrequenz-Nachrichtenkanälen auf Hochspannungsleitungen. ETZ 74 Ed. A (1953) No. 18, pp. 525–529.

35. PODSZECK, H. K.: Kapazitive Spannungswandler in Trägerfrequenz-Nachrichtenanlagen für Hochspannungsnetze. ETZ 75 Ed. A (1954) No. 19, pp. 668–671.

36. PODSZECK, H. K.: Sécurité des télécommunications dans les réseaux éléctriques en présence de bruit. CIGRE 1970, Rapport 35–01.

37. DE QUERVAIN, A.: Die Dämpfung von leitungsgerichteten Trägerfrequenzwellen durch Rauhreif. Bull. Schweiz. Elektrotechn. Ver. 42 (1951) No. 24, pp. 949 bis 953.

38. DE QUERVAIN, A.: Hochfrequenzkupplung für den Schnelldistanzschutz. Brown Boveri Mitt. 47 (1960) No. 5/6, pp. 345–352.

39. SCHUMM, E.: Fernauslösung von Hochspannungsschaltern durch Trägerfrequenzsignale. ETZ 11 Ed. B (1959) No. 12, pp. 471–475.

40. SEIDLER, E.: Ströme in Erdungsvorrichtungen und tragbaren TFH-Sperren bei Arbeiten an einsystemig abgeschalteten 220- und 380-kV-Doppelleitungen. Mitteilungen des Instituts für Energetik, Leipzig (1966) No. 81, pp. 29–41.

41. STIMMER, H.: Distanzschutz für Höchstspannungsleitungen mit gegenseitiger Schaltermitnahme über leitungsgerichtete Trägerfrequenzverbindungen. Elektrizitätswirtsch. 59 (1960) No. 17, pp. 595–598.

42. WEDEPOHL, L. M., WASLY, R. G.: Wave propagation in polyphase transmission systems. Resonance effects due to discretely bonded earth wires. Proc. Instn. Electr. Engrs. 112 (1965) No. 11, pp. 2113–2119.

43. WESTELL, E. P. L.: The Reduction of Radiation from Carrier Communication Circuits on Overhead Power Lines. J. Instn. Engrs. Austr. 24 (1952) pp. 213 to 219.

44. WOOD, T. D.: Application and operating experience of carrier communication over insulated static wires. IEEE Paper 31 CP 66–41.

Subject Index